IF YOU LOVE THIS PLANET

BY THE SAME AUTHOR

Missile Envy: The Arms Race and Nuclear War

Nuclear Madness: What You Can Do

A Desperate Passion: An Autobiography

The New Nuclear Danger:
George W. Bush's Military-Industrial Complex

Nuclear Power Is Not the Answer

War in Heaven (with Craig Eisendrath)

If You Love This Planet

A PLAN TO HEAL THE EARTH

REVISED AND UPDATED

Helen Caldicott

W. W. NORTON & COMPANY
new york · london

Printed in the United States of America

For information about permission to reproduce selections from
this book, write to Permissions, W. W. Norton & Company, Inc.,
500 Fifth Avenue, New York, NY 10110

For information about special discounts for bulk purchases, please contact
W. W. Norton Special Sales at specialsales@wwnorton.com or 800-233-4830

Manufacturing by Courier Westford
Book design by Molly Heron
Production manager: Devon Zahn

Library of Congress Cataloging-in-Publication Data

Caldicott, Helen.
If you love this planet : a plan to heal the earth / Helen Caldicott.
— Rev. and updated.
p. cm.
Includes bibliographical references and index.
ISBN 978-0-393-33302-2 (pbk.)
1. Pollution. 2. International business enterprises—Environmental
aspects. 3. Environmental protection. I. Title.
TD174.C33 2009
363.7—dc22

2009019615

W. W. Norton & Company, Inc.
500 Fifth Avenue, New York, N.Y. 10110
www.wwnorton.com

W. W. Norton & Company Ltd.
Castle House, 75/76 Wells Street, London W1T 3QT

1 2 3 4 5 6 7 8 9 0

In deep gratitude
to the persistence of
Scott Powell and Julie Enszer,
without whom this book would
never have been written

Contents

With grateful acknowledgment to Jeanette Allen, Grace Harwood, Patty Steele, Stephanie Hiebert, Erica Stern, Amy Cherry, and Scott Powell.

Introduction

The only thing necessary for the triumph of evil is for good men to do nothing.

—Edmund Burke

I first visited the United States in 1966, when I was a conservative young mother with three babies and a new medical degree. I took a job at Harvard Medical School, in the cystic fibrosis clinic of the Children's Hospital Medical Center, and worked part-time while continuing to care for my children. Because of the American films and television shows that I had grown up watching in Australia, I expected a gangster to be lurking behind every lamppost and the streets to be filled with neon lights and fast-food joints. Instead, I found the beautiful, orderly villages of New England and the quiet, reserved, but deeply caring people of Down East and Boston.

The years 1966–69, however, were years of political turbulence and violence. I witnessed the anti–Vietnam War movement, heard protest songs on the radio, watched flower children on television, and wept into my ironing as I listened to George Wald deliver an incredibly powerful and moving address on the day of the 1969 antiwar moratorium at the Massachusetts Institute of Technology.

In March of 1968, I heard the redoubtable Louis Lyons raging one night on the radio as he described the killing of Martin Luther

King, Jr., exhorting all of us to march in the streets against such iniquity. A few months later, one sultry Boston summer morning, I turned on the television to see Bobby Kennedy lying bleeding from the head on the floor of a Los Angeles hotel. I found myself screaming at the TV, "Not again!" Several days later, I drove up to the Berkshires listening to the broadcast of his funeral from St. Patrick's Cathedral, and wept as the choir of nuns burst into the "Hallelujah Chorus" from Handel's *Messiah*.

With a sense of dread, I saw Richard Nixon elected to the presidency that fall. Impelled by a growing desperation, I wrote to him about the cold war and the Anti-Ballistic Missile (ABM) Treaty, and I am sure that he used part of my letter in his inaugural address. I even wrote to Senator Edward Kennedy about my deep concern about nuclear war and the importance of the upcoming ABM Treaty, and he actually replied with a thoughtful letter, which excited this rather naïve young doctor from Down Under.

It was a thrilling time for me. Radicalized politically, I realized that democracy was a workable proposition, because the turmoil seemed to be igniting change. Anything, I thought, was possible.

In 1969, I returned to Australia and two years later successfully "took on" the French government, which was testing nuclear weapons in the atmosphere over some remote Pacific islands. The French tests violated the Partial Test Ban Treaty and created radioactive fallout in my small city of Adelaide, in South Australia. After nine months of a public education campaign warning about the medical dangers of strontium-90 and radioactive iodine in mothers' milk, 70 percent of my fellow Australians rose up and demanded that our government take action to stop the tests. The government did so, and the French were forced to test underground. I had been transformed into an activist in the United States, and my life was never to be the same again.

I have a deep regard for America as a land of excitement, change, and opportunity. It is also the country that will determine the fate of the Earth. Why do I make such a sweeping statement? Because the United States is the wealthiest and militarily the most powerful country, and because its powerful media penetrates into every corner of the world, establishing the models that most people wish to emulate. Thanks to its influence, millions of Chinese, Africans, Indians, and Latin Americans want cars, refrigerators, ice cubes in their Coke, air conditioners, and disposable packaging.

Because the US population represents only 4 percent of the Earth's people but uses 23 percent of the energy, this lifestyle is not an appropriate model for billions of other people. Such extravagant living is the leading cause of ozone depletion; global warming; toxic pollution of the air, water, and soil; and nuclear proliferation. Each US resident causes twenty to a hundred times more pollution than any Third World resident does, and rich American babies are destined to cause a thousand times more pollution than their counterparts in Bangladesh or Pakistan.[1] Canadian citizens copy the lavish lifestyle of their southern neighbors, as do most citizens of the wealthy developed countries. Many developing countries, too, as well as the dismantled Soviet Union, India, and China, are now enthusiastically embracing the "wonders" of capitalism, as their people demand this affluent lifestyle. But 6.72 billion people cannot possibly emulate the lifestyle of 300 million Americans and expect the planet to survive. Furthermore, the Earth's population may increase to 14 billion within the next century.[2]

Some eminent scientists predict that if we do not act now to reverse the cumulative effects of global pollution, species extinction, overpopulation, and the ongoing nuclear threat, it will soon—possibly within ten years—be too late for the long-term survival of most of the planet's species, perhaps even *Homo sapiens*.

My vocation is medicine, and as a physician I examine the dying planet as I do a dying patient. The Earth has a natural system of interacting homeostatic mechanisms similar to the human body's. If one system is diseased, like the ozone layer, then other systems develop abnormalities in function—the crops will die, the plankton will be damaged, and the eyes of all creatures on the planet will become diseased and vision-impaired.

We must have the tenacity and courage to examine the various disease processes afflicting our planetary home. But an accurate and meticulous diagnosis is not enough. We never cure patients by simply announcing that they are suffering from meningococcal meningitis or cancer of the bladder. Unless we are prepared to look further for the cause, or etiology, of the disease process, the patient will not be cured. Once we have elucidated the etiology, we can prescribe appropriate treatments.

Unfortunately, the etiologies of the diverse diseases afflicting our planet are complex and difficult to face when examined in fine detail. The initial, wondrous promises of capitalism and corporate free enterprise have not always led to careful and responsible management of the Earth's natural resources and treasures. And the ills of communism led in many cases to disastrous pollution and wanton neglect of nature.

In this book, I outline the diagnosis of planetary ills and then analyze the causes of these diseases. This discussion involves a critical dissection of transnational corporations and their impact on US society and the world at large. It also includes a brief analysis of the now defunct communist system relating to pollution and ecological damage.

I am trained as a physician, and my approach to problems is thus always medical. I have therefore arranged this book using the criteria one would use to diagnose and adequately treat a patient:

1. Description of signs and symptoms (Chapters 1–6)
2. Diagnosis (Chapter 7)
3. Discussion of causes of the illness (etiology) (Chapters 8–10)
4. Prescription for cures (appearing throughout the book)

The first two years of medical school, devoted to the study of basic biological sciences, are often difficult, because a tremendous amount of information must be absorbed. But they are essential to the production of well-rounded clinicians who can practice the art and science of medicine. Similarly, the first six chapters of this book are packed with basic facts about the demise of our planet, but we must have these facts at our command in order to be adequately equipped to save the Earth. As I have already said, I have a great regard for the United States and its people. I admire their resilience, creativity, and largesse. What follows in this book may strike some readers as harsh and overly critical. Yet I write out of love—and concern—for the United States and for the planet.

Students at a college in Napa Valley recently asked me, "What do you think of American people?" Having never encountered that question before in public, I searched for an honest and responsible reply. I had to say that Americans are the kindest and most caring people on Earth. They desperately want to do the right thing, but they are just not sure what that is. Is it to fight a war in the Persian Gulf and support George Bush and feel patriotic? Or is it to care for the homeless and address the domestic issues of racism and economic inequality? Or is it to save the planet?

There are no easy answers to these questions, but I will attempt to find some in this book. Let's not be afraid to look at our own society in a critical fashion, because from a caring and rigorous analysis, we will fashion a cure for the dying Earth.

If you love your country enough to cure its ills, you will be able to love and cure this planet.

• • •

With minor recent revisions, these were my optimistic words in 1991. Since that time, following the election of George Herbert Walker Bush and the first Gulf War, eight years of President Bill Clinton, and another eight years of President George W. Bush, many things have changed.

Unfortunately for the world, these presidents, their staff (excluding Vice President Al Gore), their cabinets, and the Congress were so preoccupied with the events of the day that they paid little attention to the steady beat of ominous ecological predictions emanating from the scientific literature or indeed from the first edition of this book—global warming, melting glaciers and polar ice caps, steadily rising sea levels, devastating hurricanes and cyclones, frequent and massive flooding, debilitating droughts, over a hundred million people facing starvation, extinction of many of the world's species and forests, and the onward surge of human population density.

Finally, the world can no longer ignore global warming, for it is upon us. But the solutions to this seemingly insoluble problem are at hand and will be presented in this book. They are simple, relatively cheap, and readily available. The main impediment to their implementation is the ominous and ubiquitous control of our politicians by great and powerful corporations, their money, and their lobbyists. When the politicians decide, or are finally forced, to really represent their constituents, then salvation from these human-made dilemmas can and will be attained.

IF YOU LOVE THIS PLANET

1 Ozone Depletion

There is a hole in the ozone layer. Ozone is being destroyed by chlorofluorocarbon (CFC) gases, which are like nuclear wastes in that their effects are ongoing and their life span in the ozone layer is about a hundred years. Since the hole was discovered, more than twenty-seven years ago, ozone depletion has rapidly increased. The hole (an area of almost total ozone depletion) over the Antarctic is the most severe. Ozone depletion is less severe in the stratospheric air mass of the Arctic than in the Antarctic because cold air is catalytic to ozone depletion and the Arctic is not as cold as the Antarctic.[1]

The hole varies from year to year, but the biggest losses were recorded in 2000, 2003, and 2006, when the area was more than 29 million square kilometers, the largest loss on record—an area larger than North America. Scientists predict that the ozone layer will not be able to recover to pre-1980 levels until 2075.

Before there was life on Earth, there was no ozone. Ozone protects us from the lethal consequences of solar ultraviolet (UV) light, which mutates genes and kills cells. But as evolution began, single-celled plants relatively resistant to ultraviolet light started to grow and, by the process of photosynthesis, which uses chlorophyll as a catalyst, absorbed carbon dioxide (CO_2) from the air and created oxygen (O_2).[2] The oxygen transpired from the plants and floated slowly up through the lower layers of the atmosphere (troposphere) into the upper atmosphere (stratosphere), 10–30 miles (about 15–50 kilometers) above the Earth, where it was converted to ozone (O_3) by interaction with solar ultraviolet light.[3]

Over eons of time, ozone accumulated in the stratosphere. If the ozone layer could be measured at ground-level air pressure, it would be only about an eighth of an inch (3 millimeters) thick, but at the low pressures of the thin stratospheric air it is scattered over a width of about 20 miles (35 kilometers). This gas acted as a chemical shield, keeping much of the damaging solar ultraviolet light from entering the troposphere. Because ultraviolet light kills normal living cells, no multicellular organisms (plants and animals with more than one cell) could survive before the ozone layer developed. As this protective shield thickened over time, more complex organisms were able to develop and evolve. The ozone layer protects the delicate web of life on Earth, in much the same way that sunglasses protect delicate eyes from the damaging effects of ultraviolet light. Most life would die if the ozone layer disappeared, and the Earth would return to a preevolutionary state.

In 1973, two scientists, Sherwood Rowland and Mario Molina, at the University of California at Berkeley, became worried that CFC, a human-made gas developed in 1928 by accident, could damage the ozone layer. But not until 1982 did a group of British scientists working in Antarctica verify that much of the ozone layer over the South Pole had disappeared. In 1983 and 1984, the US satellite Nimbus recorded very low levels of ozone, but these measurements were dismissed as inaccurate by computers that had been programmed to accept only normal ozone readings.[4]

In 1987 and 1989, two planes flew high into the stratosphere and managed to measure the ozone concentration, together with levels of chlorine (the element that destroys ozone). They found that 95 percent of the ozone had disappeared over Antarctica and that chlorine levels were correspondingly high. The ozone loss is apparent not just over the South Pole; since 1970, the ozone layer has decreased by 1 percent in summer and 4 percent in winter over the Northern Hemisphere in middle latitudes between 64° and 30° north.[5] This rate of depletion was approximately twice the rate that computer models had predicted.[6] Partly because of this depletion, today more than a million Americans are diagnosed annually with skin cancer. In 2007 the number of US deaths related to malignant melanoma was 7,910;[7] and one in three Caucasians will develop skin cancer in their lifetime.

The chemicals most responsible for ozone destruction belong to the family of related compounds called chlorofluorocarbons, or CFCs. These gases were used in refrigeration; in air-conditioning; in spray cans; in medical inhalers; in Styrofoam cups, trays, and packaging; as a plastic expander in foam furniture and car upholstery; and to clean computer chips.[8] Originally, it was thought that these gases were inert and harmless, did not react with other chemicals, and were safe for human contact. However, scientists failed to foresee that some CFCs remain in the atmosphere as long as 75–380 years. Once released, they rise slowly over three to five years[9] up to the stratosphere, where they interact with UV radiation, which severs the chlorine atom from the CFC molecule. This chlorine atom then splits an ozone molecule (O_3) into one oxygen molecule (O_2) and one oxygen atom (O). The chlorine is then free again to continue this cycle over many years.[10] Like an environmental Pac-Man, one chlorine atom can consume more than a hundred thousand ozone molecules over time.

Some other gases also destroy the ozone layer. The list includes carbon tetrachloride, which is used as a dry-cleaning liquid and serves as the precursor chemical for the production of CFCs; methyl chloroform, which is used for cleaning metal; methyl bromide; bromochloromethane; hydrochlorofluorocarbons (HCFCs); and halons used in fire extinguishers. HCFCs, which are now used as a substitute for CFCs, have 10–20 percent the ozone-depleting power of CFC gases.[11] Methyl bromide is a gaseous pesticide used in agriculture and in crop storage. A replacement for it has been slow to evolve, so this potent ozone-destroying pesticide is still being used.[12]

The solid rocket fuel used in the US space shuttle releases 240 tons of hydrochloric acid (HCl) into the atmosphere per launch. The chlorine atom then splits from each HCl molecule to destroy ozone molecules in the stratosphere, through which the rocket passes. In September 1991, NASA announced that it had shelved plans for many of the larger spacecraft in favor of smaller rockets and satellites, but it continues to launch an average of nine space shuttles per year.[13] To add to the problem, the US military consistently tests solid-fueled missiles for nuclear weapons delivery, and it is also engaged in launching numerous rockets to test its antiballistic missile system. Such tests exacerbate ozone depletion.[14]

As previously noted, the hole in the ozone layer over the South Pole tends to enlarge each year. This hole exists only between the months of September and October. It closes over for the rest of the year, when stratospheric ozone in the Southern Hemisphere filters down to cover the hole. This drift mechanism depletes the total ozone over the southern regions of Australia, New Zealand, South America, and Africa, particularly during their summer—November to March.[15] We Australians are sun worshipers who spend much of the summer soaking up the "rays" on our wonderful beaches; despite dire warnings from dermatologists, some people still do not use sunscreen, so the depletion of ozone caused by summertime drift down to the South Pole poses a significant threat to human health in this part of the world.

Medical and Biological Consequences of Ozone Depletion

A small amount of UV light has always penetrated the normal ozone layer, producing a certain number of skin cancers, but the incidence of skin cancer and malignant melanoma is now increasing rapidly. Some part of this increase is almost certainly due to ozone depletion. Melanoma, a dark mole that becomes malignant and is usually lethal, doubled in frequency worldwide in the two decades between 1970 and 1990.[16] In Australia, where the ozone is depleted each summer, the melanoma incidence doubled between 1980 and 1990.[17] Because of its proximity to the equator and the fact that a majority of its population is fair-skinned,

Australia experiences the world's highest incidence of melanoma.[18] The incidence of melanoma in the white population of the United States has tripled in the last twenty years.[19] It accounts for 79 percent of skin cancer deaths in the United States and ranks as the sixth most common cancer in men and seventh most common in women. In women aged twenty-five to thirty, however, it is the leading cause of cancer![20] Melanoma has risen 50 percent among young women in the United States since 1980—a trend almost certainly related to the increased use of tanning studios and exposure to the sun's rays. Among young men, however, the rate has remained steady.[21]

UV light can also damage the lens of the eye, causing cataracts and partial or complete blindness. Between twelve million and fifteen million people worldwide are blinded annually by cataracts, three million caused by UV exposure.[22] According to a study conducted by Environment Canada, if the ozone were to continue being depleted at a sustained rate of 10 percent a year, two million more people would develop cataracts globally.[23] Ultraviolet radiation also impairs the body's immune mechanism, which fights infection and cancer.[24]

Vitamin D is synthesized when UV light reacts with the skin. This vitamin is necessary for normal bone and teeth formation. A deficiency causes rickets. However, excessive UV light can induce vitamin D toxicity, causing symptoms of kidney stones and abdominal and bone pains, as calcium is mobilized from the bony structure and excreted in excessive amounts through the kidneys.

Plants, too, are very sensitive to increases in concentrations of UV light. Two-thirds of the three hundred crops that have been tested are vulnerable, including peas, beans, melons, mustard, cabbages, tomatoes, potatoes, sugar beets, and soybeans. The death of forests is very likely caused partly by increased levels of UV light, some trees being more susceptible than others. Ocean algae are also extraordinarily sensitive to UV light.[25]

It is sad that we must now cover our bodies, faces, and eyes when we go out on a beautiful day. The sun, which used to be a largely beneficial presence on which all life depended, has become, to a certain extent, our enemy.

Because of deep scientific concern about the rapidly decreasing ozone layer, representatives of many nations met in Montreal in 1987 and signed a protocol pledging to reduce CFC production by 50 percent by the year 2000.[26] Still, each 1 percent decrease in ozone could produce a 4–6 percent increase in skin cancer.[27] If the Montreal Protocol had not been enforced, however, scientists believe that by 2060, we would have had 1.5 million more cases of melanoma skin cancer than we will.

This treaty was clearly too conservative, and in June 1990, many countries agreed in London to end production and consumption of CFCs totally by 2000. The protocol was amended and further strengthened in Copenhagen in 1992, Montreal in 1997, and Beijing in 1999. Consumption of CFCs in the developed world had completely stopped by 1996, and except for HCFCs and methyl bromide, the production of all ozone-depleting chemicals was ended in the industrialized countries in 2005—a phenomenal accomplishment. By 2005, even developing countries that were allowed some leniency period were also apparently ahead of schedule. However, a thriving black-market trade in CFCs, which began in the mid-1990s as the phaseout was implemented, exists in the developing world.[28] Not surprisingly, developing countries are reluctant to phase out CFCs before the majority of their people have refrigerators, air conditioners, and so on, and the Montreal Protocol allowed them to increase their use of CFCs for a decade while they sought financial assistance to construct alternative chemical plants.

The developing nations have agreed, however, to completely eliminate CFCs by 2010 and methyl bromide by 2015.[29] Unlike CFCs, methyl bromide remains in the atmosphere for only a year.[30] The Executive Committee of the Multilateral Fund (whose task is to help signatories of the Montreal Protocol comply with the guidelines) has agreed to pay the government of China $150 million if it entirely phases out CFC production by 2010. And, with the exception of isolated cases, Latin America and the Caribbean fully reached the consumption reduction goal of 50 percent in 2005. Their next goal, which will probably not be met, is to completely phase out all CFCs by 2010.[31] Unfortunately, however, as a direct result of these treaties, stratospheric chlorine concentrations are declining but the bromine concentrations are unchanged.[32]

A study conducted by the US Environmental Protection Agency estimated that atmospheric levels of CFCs will increase more than threefold during the next hundred years, even if all CFC production ceased by the year 2000, because CFC gases will continue to leak for years from old refrigerators, air conditioners, Styrofoam cups, and plastic foam furniture.[33]

There exist excellent alternatives for refrigeration gases, such as helium and zeolites. The latter are used in Sweden for refrigeration in combination with solar power or natural gas. We could even go back to ammonia, the original refrigerant gas, which does not destroy the ozone. Because millions of people in China, India, and other huge developing countries have now or will acquire refrigerators, the need is urgent for developed countries to take responsibility and lead the way to alternative methods of refrigeration. As mentioned earlier, developed countries are to contribute $150 million annually to China to help it switch to safer alternatives.[34] If drastic steps are not taken, the ozone layer will almost certainly be depleted beyond repair. We must begin investing in companies that make safe chemicals for refrigerators. We must also be prepared not to use house or car air-conditioning and to exercise our own sweat glands during the hot weather. We can live without spray cans, plastic furniture, cold cars, and air-conditioning; and computer chips can be cleaned by other means. If the Montreal Protocol had not banned the use of CFCs during the last fifteen years, the stratospheric ozone layer would have been significantly more depleted and the result would have been a larger greenhouse effect than that caused by CO_2.

One last word about CFCs and other chemicals that deplete the ozone, including "substitutes" like HCFCs and PFCs (perfluorocarbons).[35] They not only effectively eradicate ozone, but they are as much as ten thousand to twenty thousand times more efficient as a greenhouse gas than is carbon dioxide, a product of the burning of fossil fuels. They are already responsible for 20 percent of the greenhouse warming, and their contribution is increasing. Some CFCs, however—in particular, CFC 11 and 12—caused 20 percent of the greenhouse warming, and they have been phased out.[36] However, 93 percent of CFC 114 gas is released in the United States from the uranium enrichment plant at Paducah, Kentucky.[37]

A more recent culprit has emerged on the global-warming scene: a gas called nitrogen trifluoride (NF_3) that is seventeen thousand times more potent than CO_2 as a greenhouse warmer. NF_3 is used to manufacture computers and liquid crystal display (LCD) panels on flat TV screens. It has a half-life in the atmosphere of 550 years, and the market for it has exploded because everyone wants flat TV screens. This new chemical, however, is not covered under the Kyoto Protocol, an international agreement designed to reduce atmospheric greenhouse gases. Atmospheric scientists are extremely worried about this new problem, which compounds the vexing problem of global warming.[38] Just when scientists are getting on top of the problem created by ozone-destroying human-made chemicals, another chemical culprit is rearing its ugly head.

2 Global Warming

The Earth is heating up, and the chief culprit is a gas called carbon dioxide (CO_2). Since the late nineteenth century and the beginning of the industrial revolution, the content of carbon dioxide in the air has increased by 27 percent.[1] Although this gas makes up less than 1 percent of the Earth's atmosphere, it promises to have devastating effects on the global climate over the next twenty-five to fifty years.[2] This prediction was made in 1987, and indeed it has proven to be accurate. Since then, the International Energy Agency (IEA) announced that CO_2 levels rose 15 percent between 1990 and 2005, and they are expected to increase by another 60 percent by 2030.[3] Carbon dioxide is produced when fossil fuels—coal, oil, and natural gas—burn; when trees burn; and when organic matter decays. We also exhale carbon dioxide as a waste product from our lungs, as do all other animals. Plants, on the other hand, absorb carbon dioxide through their leaves and transpire oxygen into the air.

Carbon dioxide, along with some rare human-made gases, tends to hover in the lower atmosphere, or troposphere, covering the Earth like a blanket. This layer of artificial gases behaves rather like glass in a greenhouse. It allows visible white light from the sun to enter and heat up the surface of the Earth, but the resultant heat or infrared radiation cannot pass back through the glass—the blanket of terrestrial gases. Thus, the greenhouse and the Earth heat up.

Ken Caldeira, a climate scientist based at the Carnegie Institute, says that one molecule of CO_2 generated by burning fossil fuel will, during the course of its lifetime, trap a hundred thousand times more heat than was released in producing it.[4]

In one year, 1988, humankind added 5.66 billion tons of carbon to the atmosphere by the burning of fossil fuels, and another 1 billion to 2 billion tons by deforestation and the burning of trees. Each ton of carbon produces 3.7 tons of carbon dioxide.[5] Human activities have increased the level of CO_2 in the atmosphere by 36 percent from the 1750 pre–industrial revolution level of 280 parts per million (ppm) to the present level of 385.5 ppm. This human-induced problem has increased the average global temperature by 1.4 degrees Fahrenheit (°F), or 0.8 degree Celsius (°C), with another inevitable rise over time of 1.1°F (0.6°C) because of thermal inertia—the gradual heating of the oceans, which will induce, at this stage, a 2.5°F (1.4°C) rise in global temperature. This is the highest CO_2 atmospheric concentration in the last 600,000 years, and the rate of increase has been at least ten but possibly as much as a hundred times faster than at any time in the past 420,000 years.[6] But that is not the end of the inevitable global temperature rise that we are destined to experience. If human emissions continue at the current rate until 2030, more than 0.5°F (0.3°C) will be added to the system, taking the temperature higher by 3.1°F (1.7°C). And this estimate does not include an extra 0.5°F (0.3°C) from the "albedo flip" (the phenomenon in which the dark ocean absorbs far more heat than does white ice, which reflects heat), which will occur secondary to the imminent melting of the arctic ice, raising the global temperature by the year 2030 by at least 3.6°F (2°C).[7]

The atmospheric CO_2 concentration now stands at 383.5 ppm, whereas before the industrial revolution it was 280 ppm. Scientific estimates calculate that a doubling of CO_2 concentration to 560 ppm would cause a 5.4°F (3°C) rise in global temperature.

These estimates may not be accurate, however, because they are based only on so-called fast-feedback models—which include changes in clouds, water vapor accumulation in the atmosphere, and aerosols—all of which act as heat trappers. The real warming may be much greater because of "slow-feedback" mechanisms, which include

- Ice sheet shrinkage, causing more albedo flip
- Increasing vegetation at higher latitudes, which will also absorb more solar heat because it will be darker there than the present icy tundra is
- Further greenhouse gas emissions—large quantities of methane released from permafrost melting
- Increasing ocean acidification, which itself will inhibit oceanic absorption of CO_2

NASA climate scientist James Hansen, often called "the godfather of global warming," ominously writes, "If humanity wishes to preserve a planet similar to that on which civilization developed and to which life on Earth is adapted, paleoclimate evidence and ongoing climate change suggest that CO_2 will need to be reduced from its current 385.5 ppm to at most 350 ppm. An initial 350 ppm may be achievable by phasing out coal use except where CO_2 is captured, and adopting agricultural and forestry practices that sequester carbon. If the present overshoot of this target CO_2 is not brief, there is a possibility of seeding irreversible catastrophic events."[8]

Unfortunately, according to Hansen, the prestigious Intergovernmental Panel on Climate Change (IPCC) did not take ice sheet melting and the slow-feedback mechanisms into consideration when it made its calculations on global warming in 2007, and thus it substantially underestimated subsequent rises in sea level.[9] This omission is serious because the IPCC report is considered by many, including many governments, to be the last word in climate change predictions.

Although China increased its CO_2 emissions by 8 percent in 2007, accounting for two-thirds of the growth of the year's greenhouse gas emissions, and China's total emissions were 14 percent ahead of those of the United States, Americans still produce 19.4 tons of CO_2 annually per capita—far ahead of Russia, which stands at 11.8 tons per capita; of the European Union, at 8.6 tons; of China, at 5.1 tons; and of India, at 1.8 tons. As the prices of oil and gas increase, more emphasis is being placed on coal, and 80 percent of the world's demand for coal at the present time comes from China. The United States also uses vast amounts of coal to power its industries.[10]

I lived until recently in a coastal village in New South Wales, Australia. The port of Newcastle is situated about 60 miles (100 kilometers) due north. Australia is endowed with huge amounts of coal, which is one of the country's major exports. One day the ocean horizon was littered with fifteen huge tankers waiting to enter the port of Newcastle to be loaded with coal to take to China. The image represented an archetype of the problems facing this Earth. Of course, the Australian government, although piously mouthing global-warming rhetoric, is in the lap of the coal industry, which talks endlessly about carbon capture and storage (CCS), a process that is supposed to sequester the CO_2 gas from coal burning deep underground. But from a geological perspective, it is inevitable that much of the gas will leak back into the atmosphere while at the same time poisoning and acidifying groundwater and soil. It could also trigger earthquakes.[11] The industry amazingly dubs this sequestering process "clean coal," while admitting that such a "solution" will not be operative until 2030, by which time the CO_2 derived from coal will condemn the world to irreversible warming.[12] CCS will also prove to be extremely expensive, which will inevitably remove funding from clean, readily available cheap energy such as wind and solar.

James Hansen says that because a large fraction of the CO_2 derived from fossil fuel remains in the atmosphere for a long time, one-fourth of it over several centuries, preservation of our climate requires that most of the remaining fossil fuel carbon never be allowed to be discharged into the atmosphere. And, as discussed already, because coal is the largest reservoir of carbon, exceeding that of oil and gas, coal emissions must be reduced and phased out by 2030.[13]

Because of this warming, the Arctic sea-ice sheet is melting much faster than predicted—one hundred years ahead of schedule, declining 22 percent in the last two years alone. In fact, some scientists predict that the Arctic could be ice-free in summertime as soon as 2010.[14] As the white ice turns to black sea, the heat from the sun is absorbed into the blackness instead of being reflected back into space. But the increase in sea temperature resulting from this albedo flip could then induce an unexpectedly rapid melting of the Greenland ice mass—possibly three times faster than presently predicted. In fact, the Greenland ice is already disintegrating in ways that were not foreseen: the surface ice is melting; deep rivers are developing, which are cutting through the ice sheet; and lubricating water is accumulating between the rock base and the ice mass itself, helping the ice sheet to move toward the sea. Glaciers on Greenland are also moving faster into the sea, and large chunks of ice are breaking off and lurching forward before grinding to a halt. These massive movements of ice are inducing earthquake-like tremors that are now becoming an ominous, common occurrence.[15] Melting of the Greenland ice could raise sea levels nearly 25 feet (7.3 meters).

Rapid, deleterious change is occurring in all the arctic systems, seriously affecting the atmosphere and oceans; the sea ice and ice sheets; the snow and permafrost; food webs; ecosystems; species survivals and extinctions; and last but not least, human societies that have lived and survived in these northern geographic areas for eons.[16] In fact, so serious is this rapid change in the arctic ice sheet that sea lanes are opening up, allowing access to previously unreachable deposits of oil and gas (approximately 90 billion gallons [about 340 billion liters] of oil and 1,669 trillion cubic feet [about 47 trillion cubic meters] of natural gas), which is inciting a race between the United States, Canada, and Russia to lay claim to them, all at a time when it is imperative that we stop burning oil and gas as soon as is feasible.[17]

Next we need to look at the Antarctic, where there is also a massive quantity of ice. Although the East Antarctic ice sheet is larger than the West Antarctic ice sheet, the latter is considered more vulnerable to global warming, and its disintegration could cause sea levels to rise another 16 feet (5 meters) by century's end. In March 2002, the West Antarctic's Larsen B ice shelf, which had been stable at a thickness of some 650 feet (200 meters) for twelve thousand years, collapsed. These events are unpredictable and can happen suddenly and much faster than humankind has estimated.[18] These sudden, unexpected climatic events are called "tipping points."

James Hansen says there is "already enough carbon in Earth's atmosphere to ensure that sea levels will rise several feet (meters) in coming decades." In fact, Hansen predicts that if the world continues to operate in a "business as usual" mode, doing little to avert greenhouse emissions, the rise in sea level is likely to be about 16 feet (5 meters) this century.[19]

Just with moderate disintegration of these ice sheets, the human consequences would be catastrophic. According to Sir Nicholas Stern, the British economist who wrote a prestigious report on climate change, over two hundred million people reside in coastal floodplains, with 0.8 million square miles (2 million square kilometers) of land and $1 trillion of assets less than 3 feet (about 1 meter) above current sea level. Thirty-five million people in Bangladesh (one-fourth of the population) live on the coastal floodplain, and twenty-two of the world's largest cities are threatened by flooding, including Tokyo, New York, Venice, Miami, London, St. Petersburg, Hong Kong, and Buenos Aires.[20]

However, because about one-third of the human population lives within about 40 miles (60 kilometers) of the sea, millions—even billions—of people could either be killed by floods or catastrophic storms or be forced to migrate to higher elevations, thereby severely dislocating other urban and rural populations. These ecological refugees will create chaos as they move into established rural areas, towns, and cities. Food production will already have been disrupted by the change in climate, and a redistribution of the scarce remaining resources will probably not happen.[21]

More than two billion people depend on fresh groundwater, and long before rising seas actually inundate the land, aquifers will become contaminated with salt, imperiling the water supplies of dozens of cities and millions of people.[22] Of course, this problem will be exacerbated by falling water tables secondary to global warming and the ever-increasing use of water by the world's growing urbanized populations.[23]

Landslides in the European Alps are serious as permafrost melts away and glaciers melt, and the Kilimanjaro ice sheet in Africa, which has existed for eleven thousand years, is disappearing, exhibiting 80 percent loss in the last hundred years. As glaciers disappear globally, large down-river populations, which depend on the glacial freshwater, will be severely threatened. The Hadley Centre, which focuses on the scientific issues associated with climate change in Britain, says that a global warming of just 1°C would eliminate freshwater from one-third of the land's surface by the year 2100.[24]

Adding to the catastrophes of rising sea levels, storm surges will further damage densely populated river deltas and coastal infrastructures, including ports, sewers, railways, water services, and electricity transmission, destroying domestic and industrial structures.[25] A report issued in June 2008 by the US Climate Change Science Program (CCSP) warned that unpredictable and severe weather events would exact a heavy toll. This is already occurring, as the rise in the ferocity and frequency of hurricanes hitting the North American continent attests. The CCSP report indicated that in the last two decades, there have been fewer cold snaps than in any other ten-year period on record, and in six of the last ten years we have experienced average temperatures falling within the top 10 percent of all recorded years. The CCSP models predicted that intense precipitation events that used to occur every twenty years will occur every five years now, and that the US Southwest will likely face even more intense droughts.[26]

For Australia, which is already facing a severe drought, a report written by the economist Ross Garnaut predicts dire events unless the world community faces up to its responsibilities to end CO_2 emissions. There could be as many as ninety-five hundred heat wave–related deaths per year, an end to agriculture in the Murray-Darling basin (our main river system and the heart of Australia's agriculture), exposure of 5.5 million people to the dengue virus, destruction of the wonderful Great Barrier Reef as the coral is bleached by high water temperature and ocean acidification, a 4.8 percent reduction in the GDP, and political instability in neighboring countries.[27]

Another report, issued at the same time by the Australian Bureau of Meteorology and CSIRO (the Australian Commonwealth Scientific and Industrial Research Organisation), predicts that Australia will suffer severe heat waves and droughts over wider stretches of the country every year for the next three decades.[28]

Interestingly, the cynical Pentagon is already in on the act. It predicts that climate change will be a greater global threat than terrorism. The Pentagon says that wars will no longer be fought over land or religion, but over sheer survival. Major waterways such as the Amazon, Danube, and Nile will become battlegrounds for water rights. Millions are expected to die of famine, especially in the subtropical regions. Nuclear arms will be produced by Japan, South and North Korea, Germany, Iran, and Egypt; and Israel, China, India, and Pakistan are expected to use the bomb.[29] Of course, this is a perfect recipe for annihilation. Instead of judiciously and sagaciously trying to organize the world for survival in an age of great impending catastrophes, our wonderful and well-financed warriors see global warming as a legitimate and rational reason for ever more wars.

To put global warming into perspective, however, carbon dioxide accounts for only half of the greenhouse effect. Other gases, the so-called trace gases, which are present in minute concentrations, are much more efficient at trapping heat. We already discussed the fact that CFCs are ten thousand to twenty thousand times more efficient than carbon dioxide as they rise up through the troposphere. Methane is also a very efficient heat trapper (twenty times more effective than carbon dioxide) and is released at the rate of about 25 gallons (100 liters) per day from the intestine of a single grain-fed cow. Australia's cows make an annual contribution to global heating equivalent to the burning of 13 million tons of black coal (about half the coal used in Australia per year). An outdoor test conducted on cows that graze on pasture in Australia showed that each cow emits more than 90 gallons (350 liters) of methane per day.[30] The scientists Ralph Laby and Ruth Ellis, from CSIRO, developed a slow-release capsule that diminishes by 20 percent the production of methane by bacteria in the rumen of a cow. (Methane is also a wonderful gas for heating and lighting houses; for example, Laby and Ellis estimated that two cows produce enough methane to heat and light an average house!)[31] In fact, in the United States, cattle-produced methane and manure has a global-warming effect equivalent to that of thirty-three million automobiles.[32] Additional sources of methane are garbage dumps, rice paddies, and termites.

Another greenhouse gas, nitrous oxide, is a component of car and power plant exhausts, of chemical nitrogenous fertilizers, and of bacterial action in heated, denuded soil. Nitrous oxide is three hundred times more efficient than CO_2 as a heat trapper, and it has contributed about one-tenth as much to global warming as carbon dioxide has. Nitrous oxide in the atmosphere has increased by 18 percent over pre–industrial revolution levels; and methane, by 100 percent.[33]

A report by the World Wide Fund for *Nature*, published in August 1991, stated that carbon dioxide emissions from aircraft flying at altitudes of about 6–7 miles (10–12 kilometers) account for 1.3 percent of the global warming. A more recent (2006) report said that now aircraft account for 4 percent of the world's carbon dioxide emissions and scientists predict that by 2030, aircraft will account for one-fourth of the total global CO_2 emissions.[34] However, aircraft also emit nitrous oxide gas, which is an extremely efficient heat trapper at the height at which planes fly, and this gas itself may increase global warming by 5–40 percent.[35] In summary, at the moment the total warming effect of aircraft emissions is 2.7 times as great as the effect of the same quantity of emissions at ground level,[36] and air travel provides the fastest-growing source of global-warming CO_2 emissions, increasing almost 5 percent a year, according to the European Environment Agency.[37]

In December 2006, the US Air Transport Association said that US airlines had improved their jet fuel efficiency by 70 percent in the previous thirty years. Yet, on average, new jet engines emit more nitrous oxides than do older engines. Boeing is researching ways to reduce emissions by combining hydrogen and oxygen as fuel, but it is at least ten years away from producing a prototype. NASA is also developing methods for the Boeing 737 and the Airbus A320 to burn 25 percent less fuel and to reduce nitrous oxide emissions by 80 percent, hoping to complete this work by 2018.[38]

Eighty percent of the world's carbon dioxide emissions have been produced since 1940.[39] It was predicted in 1989 that within fifty years, the "effective carbon dioxide concentration" (CO_2 and trace gases) will probably be twice that of preindustrial levels, raising global temperatures 2.7°F–10°F (1.5°C–5.5°C).[40] In the last two decades, carbon dioxide has increased on average 1.5 ppm a year. But between 2002 and 2003, CO_2 rose 2 ppm. If current trends continue, temperatures could rise between about 1.8°F and 7.2°F (1°C–4°C) within the next twenty to thirty years.[41]

Such rapid change in climatic conditions has never occurred before in human history. If global heating were at the higher predicted level, it could then match the 9°F (5°C) warming associated with the end of the last ice age, eighteen thousand years ago. But this change would take place ten to a hundred times faster.[42] And at present temperatures, a 9°F (5°C) increase would cause global temperatures to be higher than at any other time during the last two million years. These predictions were made by the eminent climatologist Stephen Schneider in 1989, and they match present predictions.

What will happen to the Earth? Let's look at a worst-case scenario. Changes in climate could have devastating consequences in the tropical forests and food-growing areas of the world, causing the extinction of many plant and animal species over only a few years, in evolutionary terms. Dust bowls could develop in the wheat belt of the United States, creating a landscape like that described in *The Grapes of Wrath*, and the productive corn and wheat belt might migrate north into Canada and into the Russian republics.[43]

The futures markets that make money out of gambling with predictions, were even speculating on impending global disaster in 1989, predicting that productive banana and pineapple plantations would develop in the middle of arid Australia and that cyclones, tidal waves, and floods would almost certainly affect temperate areas of the world, which had previously been immune to such catastrophes.[44]

In 2006, Sir Nicholas Stern predicted a 5–10 percent increase in hurricanes and strong winds, which will increase sea levels and double annual damage costs in the United States.[45] Sir John Holmes, the United Nations emergency relief coordinator, warned that twelve of the thirteen major relief operations in 2007 were climate-related and that this situation amounts to a climate change "mega disaster."[46] In 2008, however, Stern revised his original estimates, saying that he had "badly under-estimated the degree of damages and risks of climate change," advocating a 50 percent reduction in greenhouse gases by 2050—a goal that would require a 90 percent reduction by the United States.[47]

After two category 5 hurricanes hit in the same season during 2007—Hurricane Dean in Mexico and Hurricane Felix in Nicaragua—the insurance industries experienced financial distress. With record losses and payouts far exceeding premiums in recent years, insurance rates are rising and companies are moving out of US coastal towns. The UN predicts that climate-change disasters will cost the world's financial centers as much as $150 billion annually within a decade.[48]

To make matters worse, the world is now witnessing hurricanes, tornadoes, and cyclones of unprecedented power and frequency. Hurricane Katrina on August 29, 2005, killed 1,400 people in the United States and caused $200 billion in damage.[49] About a quarter of a million people fled New Orleans, and this flight by Katrina's victims became the first massive movement of refugees secondary to a hurricane. About one-fourth of them have settled elsewhere and will not return.[50] Then, Cyclone Nargis hit Burma on May 2, 2008, resulting in at least 130,000 deaths and the displacement of two million to three million people, 1.5 million of whom were severely affected.[51] To make matters worse, the secretive Burmese government has virtually prevented aid from reaching the devastated populations, and government workers have been found distributing foreign aid and goods to their friends instead of to those in need.

Sea levels have already risen significantly. From 1961 to 2003 the average sea level rose a little over a sixteenth of an inch (2 millimeters, mm) a year; between 1993 and 2003, about an eighth of an inch (3 mm) a year.[52] Sea levels had already risen about 4–8 inches (10–20 centimeters, cm) in the previous hundred years.[53] Future predictions vary—the IPCC predicts that sea levels will rise between 4 and 35 inches (9–88 cm) by 2100, but their predictions are almost certainly underestimated, and these levels may rise about 15 feet (5 meters) or more.[54] Recent research by an Australian and American team found that the rate of warming in the upper layers of the oceans between 1961 and 2003 was 50 percent higher than had been estimated by the 2007 United Nations Intergovernmental Panel on Climate Change.[55] Some of the dynamics of the observed increase in sea level can be explained by the fact that as water heats, it expands.

As ocean levels rise, rivers, lakes, and estuaries will have their courses and boundaries changed forever. These changes, of course, will disturb the hatching habitats of millions of fish.[56] The aquatic food chain will be threatened because the organisms forming the base of the pyramid of the ocean food chain—algae and plankton—are being seriously affected. These ubiquitous single-celled organisms are food for primitive life-forms and are themselves consumed by more evolved species of fish. Some forms of algae and plankton are threatened by rising sea temperatures, and many are also extremely sensitive to UV light—so sensitive that some species have a built-in mechanism enabling them to dive from the surface to lower depths at midday, when the UV light is most powerful, to escape the lethal radiation. Therefore, as the temperature rises and as the ozone diminishes, this essential element of the food chain will be jeopardized. In fact, marine phytoplankton are being dramatically affected already by rising sea temperatures. Phytoplankton account for nearly half the photosynthesis of all the plants on Earth. As sea temperatures rise, phytoplankton production will decrease, and therefore the rate of CO_2 absorption will diminish. Loss of these organisms could lead to a "positive-feedback cycle," meaning that CO_2 levels in the atmosphere will increase.[57]

Moreover, plankton and algae, together with trees and plants, are nature's biological traps for elemental carbon from atmospheric carbon dioxide. Sea plants trap 41 percent, and land plants trap 59 percent. Higher concentrations of atmospheric carbon dioxide will actually promote the growth of algae because CO_2 acts as a photosynthetic fertilizer. But if algae are threatened by global warming and an increased concentration of UV light secondary to ozone depletion, this hypothetical fertilizer effect will become irrelevant. By increasing the atmospheric concentration of carbon dioxide from human-made sources, we are also threatening the survival of trees, plants, algae, and plankton—the very organisms that help reverse global warming.

Forests, too, are terribly vulnerable to climatic change and ozone destruction. Because temperature changes are already occurring at alarming rates, specific tree species will not have thousands of years to migrate to latitudes better suited to their survival, as they did at the end of the last ice age. When the ice cap slowly retreated northward at that time, the spruce and fir forests moved from the area of the United States into Canada at the rate of 0.6 mile (1 kilometer) per year. Although some plants that adapt rapidly will thrive under changed circumstances, most forests will die, and along with them will go many animal and bird species.[58]

Although sudden global warming will kill large numbers of trees, increased carbon dioxide concentrations will also stimulate the growth of the trees that remain, because CO_2 is a plant food for trees as well as algae, acting as a fertilizer.[59] Therefore, as forests become extinct in the unusually hot climate, some food crops and surviving trees will grow bigger and taller. Unfortunately, many weeds are even more responsive to high carbon dioxide levels than are crop plants, and they will almost certainly create adverse competition.[60]

Another factor to consider in this rather dire biological scenario is that faster-growing crops use more soil nutrients. Hence, more artificial fertilizer will be needed, and since electricity is required for its production, more carbon dioxide will be added to the air. But nitrogen-containing fertilizers themselves release the greenhouse gas nitrous oxide into the air.[61] In addition, as soil heats, vegetable matter decays faster, releasing more carbon dioxide. These are just a few of the interdependent and variable effects of global warming that are so difficult to calculate.

When forests are destroyed by greenhouse and ozone deforestation, or by chainsaw and bulldozer, massive quantities of rich topsoil are being lost forever as floods and erosion wash it out to sea. Downstream waterways overflow their banks as rain pours off the denuded high ground, and when the floods subside, the once deep rivers will be clogged with silt from the eroded topsoil. Large dams designed for predictable rainfalls will collapse and drown downstream populations, and associated hydroelectric facilities will be destroyed.

Massive rainfall in various parts of the world will continue to have unnerving impacts. As I write, severe flooding has disrupted many parts of Iowa, Wisconsin, Illinois, and Missouri. Rivers, including the mighty Mississippi, are peaking at almost unprecedented levels, seen only once in the last five hundred years. Large areas of corn and grain crops have been flooded and washed away. This is a serious catastrophe because (1) much of the world is experiencing a severe food shortage as the price of oil rises precipitously, (2) grain is currently being used for biofuels, (3) the stock market is speculating on "commodities" such as food, and (4) droughts in Australia and elsewhere are seriously cutting into the export trade. Some parts of Iowa have experienced more than the average annual rainfall in a single storm or two. Not only is precious topsoil washing away, but nitrogenous fertilizer is leaching away with the soil. This dark, nutrient-laden flood will soon enter the Gulf of Mexico, where it will add to the large "dead zone" where no fish or aquatic life exists. Tragically, when these floods occurred, a horrendous tornado also hit the western part of Iowa, killing four Boy Scouts as they camped.[62]

At the same time, massive flooding is also disrupting rescue efforts in the earthquake-ridden area of southern China—where the Yangtze and Pearl rivers are overflowing their banks—following the worst storms there in fifty years. Large swaths of farmland are flooded, the economic costs total $1.5 billion, and almost eighteen million people have been affected.[63]

In other parts of the world, decreased rainfall will reduce stream runoff. For example, a rise in temperature of several degrees Celsius could deplete water levels in the Colorado River, causing severe distress for all communities that depend on the river for irrigation, gardening, drinking water, and so forth. The water quality will also suffer, because decreased volumes will not adequately dilute toxic wastes, urban runoff, and sewage from towns and industry.[64] "Until April 1991, when rain began to fall again in some quantity, California experienced a severe five-year drought, whose impact was rapidly becoming critical. After this April rainfall, the California drought continued unabated. That may be an omen of worse to come." I wrote these words prophetically in 1991. What has happened since? California produces a larger quantity of greenhouse gas emissions than all other forty-nine states combined—roughly 500 million metric tons annually. If current trends continue, experts predict that Los Angeles will, on average, experience a hundred or more days a year above 90°F (32°C)—a situation that will obviously further deplete rivers and reservoirs.[65]

In June of 2006, 45 percent of the United States experienced a "moderate-to-extreme drought." In only one summer, fifty thousand wildfires burned 3 million acres (1.2 million hectares) across the nation, adding millions of tons of CO_2 to global warming, as well as killing many indigenous birds and animals.[66] As the forests dry and turn to tinder, the hot air and wind catalyzes firestorms, which aggravate global warming. It is then a vicious circle. And as I write in July 2008, California is facing disastrous wildfires, burning on 1,400 fronts, that were started by lightning strikes on tinder-dry undergrowth.[67]

The US drought from 1999 to 2006 ranks as the third worst, behind the droughts of the 1930s and '50s. Some farmers moved their cattle to greener pastures, if there were any. Many others were forced to sell their herds.[68]

Australia is now experiencing its most severe drought in a hundred years. The paddocks and fields are bereft of grass, and the soil is exposed to winds that scoop it up in violent dust storms. Farmers are desperate, many leaving their properties, and there is no guarantee at this time that decent rains will fall in the near future. The great River Murray, the largest river in Australia, is drying up;

Lakes Alexandrina and Albert at the mouth of the Murray River—havens for thousands of flocks of waterbirds—are almost nonexistent, and the drying mud is giving off noxious sulfurous fumes. Adelaide, the capital city of South Australia, which relies on the Murray, may be without water within the next twelve months.

In Europe, Spain lost $2 billion in agriculture in 2005 because of severe droughts. The Spanish drought was the worst experienced in half a century.[69] Large areas of Africa are also experiencing severe drought, the results of which are reflected in the nightly news bulletins reporting thousands of malnourished and starving refugees vainly searching for food and water.

As the planet warms, cities will become heat traps. For instance, Washington DC at present suffers one day per year over 100°F (38°C) and thirty-five days over 90°F (32°C). By the year 2050, these days could number twelve and eighty-five, respectively. In that case, many very young and old and infirm persons would die from heat stress. There would be a great and universal temptation to turn on air conditioners, which, of course, use global-warming HCFCs and electricity, the generation of which produces more carbon dioxide, or nuclear waste if nuclear electricity is used. People will thus be in a catch-22 situation—damned if they do and damned if they don't. The average summer temperature in the eastern United States currently ranges from about 80°F to 86°F (27°C–30°C). A new research study by NASA suggests, however, that by the 2080s the average summer temperature could be in the 90s Fahrenheit (low- to mid-30s Celsius).[70]

Is the human race smart enough, and does it have the intellectual, psychological, and emotional resources and stamina, to reverse this ongoing disastrous situation?

3 The Politics of Environmental Degradation

How did the problem of atmospheric degradation become so alarming, and what are the solutions?

When CFC (chlorofluorocarbon) was first concocted, in 1928, nobody understood its impacts on the complexities of atmospheric chemistry because it was an inert gas that did not react with other chemicals. During subsequent decades, scientists really believed that chlorofluorocarbons were ideal for refrigeration, air conditioners, plastic expanders, spray cans, and cleaners for silicon chips.[1] Industry became so heavily invested in production that it was difficult to cut back, despite the serious consequences of failing to do so. At the 1990 London International Conference on Ozone, a group of enthusiastic young Australians made presentations begging for a safe future. One middle-aged conference participant approached the teenagers and said, "Look, my attitude is that if you are on the *Titanic*, you may as well have the best berth." Such cynical acceptance of the planet's demise is suicidal and demonstrates a total lack of a sense of responsibility for the next generation. In fact, since the ratification of the 1987 Montreal Protocol to reduce CFC production, some major chemical manufacturers have even attempted to deny that these chemicals destroy ozone in the atmosphere. At international meetings, there has been continual intense lobbying to protect the interests of industry.[2]

In the early years of the industrial revolution, no person could have predicted the atmospheric havoc that the internal combustion engine and coal-fired plants would wreak. Even during the 1930s and '40s, when General Motors, Standard Oil, Phillips Petroleum, Firestone Tire and Rubber, and Mack Manufacturing (the big-truck maker) bought up and destroyed the excellent mass-transit systems of Los Angeles, San Francisco, and many other large US cities in order to induce total societal dependence on the automobile, global warming was a vague future threat.[3] These companies were subsequently indicted and convicted of violating the Sherman Antitrust Act. But it was too late; these actions of vandalism were irreversible.

Now that we understand the coming disaster, however, we are in a position to act. But in order to do so, we must be willing to face several unpleasant facts.

Fact Number One: US Energy Use Is Disproportionately High

The United States, constituting only 4 percent of the Earth's population, is responsible for 23 percent of the world's output of carbon dioxide. It uses 35 percent of the world's transportation energy, and an average-size tank of gasoline produces 300–400 pounds (136–181 kilograms) of carbon dioxide when burned.[4]

In the 1970s there were 200 million cars in the world. By 2006 there were 850 million cars, and this number is expected to double by 2030. And on average, urban car travel uses twice as much energy as travel by bus, 3.7 times as much as travel by light rail or tram, and 6.6 times as much as travel by electric train.[5]

For additional perspective on our use of energy resources, let's compare the situation in the United States with that in China. Relevant facts for both countries follow.

RELEVANT FACTS ABOUT THE UNITED STATES

- On average, 40,000 pounds (about 18,000 kilograms) of CO_2 is emitted each year by every US citizen (more than China, India, and Japan per capita combined).
- The United States consumes eight hundred million trees per year, or 50 million pounds (about 22.5 million kilograms) of paper.
- The United States uses over 200 million gallons (about 750 million liters) of gas a day.
- US cars contribute 56 percent of all air pollution.
- The United States consumes 25 percent of the world's oil production, and 16 percent of the world's oil is used in US cars.
- There are over twenty million gas-guzzling four-wheel-drive vehicles on US roads today.
- Fifty-two million new cars were made annually in 2007, ten and a half million in the United States.[6]
- Only 1 percent of Americans use public transport.
- The average American produces 1,905 pounds (864 kilograms) of waste per year—three times more than the average Italian.[7]
- Of America's electricity, 50.3 percent was generated by one hundred coal-fired plants in January 2008, with approximately 20 percent from nuclear power and 20 percent from natural gas. Petroleum generated only 1.3 percent of US electricity.[8]

RELEVANT FACTS ABOUT CHINA

In China in 1991, there were three hundred million bicycles, and only one person in seventy-four thousand owned a car. Each year, three times more bicycles than cars were produced. Domestic sales of bicycles in China exceeded the sale of cars worldwide in 1987 by thirty-seven million.[9] Only seventeen years later, in 2008, as China emulates the United States and Europe, vociferously and without reserve joining the capitalistic ethos, the situation is quite different:

- China has a population of over 1.3 billion people.[10]
- Though the Chinese government does not publish carbon emission reports, foreign analysts believe it is now the biggest greenhouse gas emitter, because of its massive consumption

of coal, gas, and oil. Scientists predict that by 2030, China's emissions will be twice as large as the emissions of all the OECD (Organisation for Economic Co-operation and Development) countries combined.[11]

- Bicycles are now forbidden on many streets in big Chinese cities, and ever-more car lanes are under construction.[12]
- China uses three times more energy per dollar of its GDP than the global average, and 4.7 times more than the United States.[13]
- Forty new Chinese airports are planned for construction within the next thirty years.[14]
- Eighty percent of China's electricity is generated from coal, and China is building on average one new coal-fired plant every week.
- China has very little in oil or gas reserves, but it has 13 percent of the world's coal reserves—enough to sustain its economic growth for the next hundred years.[15]
- China has contributed less than 8 percent of the total emissions of CO_2 from energy use since 1850 (the United States is responsible for 29 percent; and Europe, for 27 percent).[16]
- China's prime minister, Wen Jiabao, said that his country is prepared to limit global-warming emissions and to replace the outdated Kyoto Protocol provisions, which expire in 2012. Plans include the following:
  - Building nine high-speed passenger railway lines.
  - Restricting car emissions in 2007.
  - Creating more wind and small-scale hydroelectric plants in the next fifteen years, capable of generating 40,000 gigawatts—this plan will account for 4 percent of China's power supply by 2020. Partial funding will come from the US government.[17] This may change under President Obama.

Fact Number Two: Urgent Response Is Necessary to Curb Greenhouse Warming

In order to reduce carbon dioxide production, cars must be made extremely fuel-efficient, and some computer models and prototype automobiles can indeed achieve 60–120 miles per gallon (mpg) by means of lightweight materials and better design.[18] But these techniques are rarely employed. In 1987, for instance, US car manufacturers dropped most of their research on fuel-efficient cars; and in 1986 the fuel-efficient standard, or minimum mileage, in the States was only 26 mpg. By 1991, it was still only 27.5 mpg, and the first Bush administration steadfastly resisted any move to increase fuel efficiency in cars.[19] In fact, President George H.W. Bush's energy plan of 1991 barely dealt with these issues, and it totally failed to address mass transportation.

During the intervening years, energy efficiency of US cars was not improved under either Clinton or George W. Bush. In fact the sales of four-wheel-drive SUVs (in particular, Ford's F-150) in 2007 showed that the SUV was the top-selling vehicle in the United States—a position it had held for the previous twenty-six years. Although overall sales of big pickups have declined since then by 6 percent because of higher fuel prices and a slowdown in housing starts, they are still a top-selling item in a declining market.[20] Even the hideous Hummers first designed for Desert Storm are still popular, and many can be seen on the streets of New York. Its seems that the more ominous the signs of global oil depletion are, the more determined some people are to deny reality—as if the future will be the same as the past. However, the market for these large sport utility vehicles plummeted after the price of gasoline reached $4 a gallon in the second half of 2008. And this drop in the SUV market was exacerbated when the three US auto companies came under severe financial strain as car sales in general declined in September 2008 because a reeling economy scared potential buyers and many buyers were unable to get loans.[21]

Clearly, it is more efficient to transport hundreds of bodies in one train than hundreds of single bodies in hundreds of cars. Furthermore, the construction of sleek, state-of-the-art trains would have constructively reemployed the one in eight people in California who had been employed producing weapons of mass destruction.[22] Far more people could have worked in this wonderful new civilian industry at the end of the cold war than in the obsolete weapons industry, because the military sector is capital-intensive, whereas the civilian sector is labor-intensive. However, President Clinton allowed the military-industrial complex to proceed apace in many areas, as if the cold war had not ended; and then George W. Bush, riding high on the 9/11 tragedy and aided and abetted by Donald Rumsfeld and cronies, ramped up the US military budget to unprecedented heights. Unfortunately, these wonderful opportunities that the termination of the cold war presented were never implemented. One of the main reasons that the "peace dividend"—the opportunity that peace between the US and Russia provided—was never pragmatically implemented was because of the power, might, and lobbying influence of the military-industrial complex in the US political sphere. But the rapid construction of an efficient mass-transit system in the United States is now an imperative. The corporations that accept this challenge could become the world's leading producers of global mass-transit systems, earning huge profits while simultaneously helping to save the planet. No doubt, President Obama will make mass transit one of his top priorities as he attempts to counter global warming and to put millions of people back to work.

Now we must address another impending problem impinging on the global economy, particularly on transportation: peak oil. "Peak oil" is defined as the time when the global rate of extraction of oil will reach a maximum before it inevitably declines. Note that oil is not used to generate electricity, so when the nuclear industry and politicians claim that nuclear power is the answer to the ever-diminishing supplies of oil, they are obfuscating the truth. Oil is used primarily for transportation, and as such it is the foremost energy resource for the world. There have been many recent estimates of when global peak oil will occur, but it is possible that this date was actually reached in May 2005, when global production rates of crude oil reached 74.2 million barrels a day; production has been falling ever since.[23]

As I write, in June 2008, the price of a barrel of oil recently reached $140, seriously disturbing Wall Street and the global stock market, and nobody knows where it will end. Indigenous US oil production peaked in 1970; consequently, the United States is now the world's largest importer of oil. Of the fifty-one oil-producing nations, twenty-seven are reporting declines in production.[24]

A study commissioned by the US Department of Energy and published in 2005 (the Hirsch Report) forecast unprecedented social, economic, and political impacts in the United States if efforts were not immediately commenced on a "crash program" scale, to be implemented at least ten years in advance of peak oil, so that demand for oil would be reduced together with the initiation of large-scale production of alternative fuels.[25] The downward trend in oil production since May 2005 seems to suggest that the Hirsch Report may have been ten years too late!

The entire world economy is at risk because high prices and real shortages of available oil will dramatically damage national economies:[26]

- The entire global transport system and the "globalization" economic model—passenger and freight planes, mass transportation, shipping, trucking, cars, diesel passenger and freight trains—is based almost entirely on oil. Obviously, then, the cost of travel and prices for all commodities will soar.
- Agriculture, as it has been constructed over the past century—the production of fertilizers and pesticides, and the planting and harvesting of food—is also based almost entirely on oil. In fact, about 10 calories of fossil fuel is used to produce each single calorie of food. The global transportation system and the trading of food are totally dependent on oil. Amazingly, the average distance traversed by food in the United States is over 1,500 miles (about 2,400 kilometers). Obviously, the shortage of oil will translate into ever-higher food prices and inevitable shortages.
- Plastics and most chemicals are made from oil. The three main compounds butadiene, ethylene, and propylene used to create plastics, clothing, toys, packaging, solvents, disinfectants, pesticides, coolants, antifreezes, lubricants, and so on are all gleaned from oil. The entire petrochemical industry will be severely affected.

Oil shortages will inevitably lead to domestic, national, and international tensions and conflict. As Richard Heinberg says in his excellent paper "The View from Oil's Peak," "the crisis of oil will not be solved in days, weeks or even years. Decades will be required to re-engineer modern economies to function with a perpetually declining oil supply."[27]

This crisis will not be solved by the many harebrained schemes that are currently under consideration to develop the gasification of coal or the direct liquefaction of coal to produce liquid transportation fuels. Not only do these extremely expensive technologies require massive amounts of water on a drying planet, but they also produce huge amounts of CO_2, which would be catastrophic to the global climate. China is investing billions in the development of these technologies on massive scales, but they will depend on carbon capture and storage (CCS) to remove the CO_2 from the atmosphere. As noted earlier, CCS technology is decades away from implementation, and it may never be scientifically or practically successful.[28]

What, then, are some of the solutions?

- Electric transportation, as described next, for both cars and mass transit
- Bicycles
- Locally grown food using tried and true permaculture and organic gardening techniques
- Eliminating the unnecessary packaging of food and many other items
- Drastically reducing plastic consumption and mass chemical production

Hybrid cars that use both electricity from a battery and gasoline to operate the motor are becoming de rigueur. Toyota has now sold over a million hybrid cars. The Prius hybrid uses 4.5 liters per 100 kilometers (making the average mileage 64 miles per gallon). The average car does only 20–25 mpg, and SUVs do 10–17 mpg. The logical question to ask is, Why are SUVs still so popular, and why are they still allowed to be made? They should be banned by law, if we are committed to saving the planet. Plug-in hybrids need to be powered by solar electricity, not by coal- or nuclear-generated power. Such technology is already in use: in 2002 the US Navy installed a massive solar-powered parking lot near San Diego to power plug-in hybrids.[29] Google also has a trial system in place in Silicon Valley using plug-in hybrids.

Other wonderful new inventions are on the horizon: all-electric cars powered solely by solar power. These cars are described as "vehicle-to-grid" (abbreviated "V2G").[30] With the help of newly developed lithium-ion batteries that can be recharged ten thousand to fifteen thousand times, these cars can be plugged into a solar-powered parking lot, charged up during the day, driven home at night, and then plugged into the house to provide power for the residence at night. In fact, there is enough area on the roofs of all the parking lots in the United States to power the country's entire fleet of cars by solar energy and then to add this electricity to the grid to power homes at night.

The installed power of engines in cars and light trucks in the United States is more than one order of magnitude higher than that of the entire US electric power system. Ten million solar-powered vehicles could supply a standby capacity of 100,000 megawatts, the equivalent of power generated by a hundred nuclear power plants.[31] This technology is ready to go. Now the politicians must mandate that all cars be solar-powered and allocate the required funding to subsidize and support this industry until it gets off the ground.

Cars can also be fueled with direct solar energy. In 1990, Australia held an international solar car race across the country. The cars were equipped with photovoltaic panels on their roofs, and they achieved the acceptable speed of approximately 60 miles per hour, or mph (about 100 kilometers per hour, or kph). They were slow to accelerate, but who needs cars that go from 0 to 90 mph (0–145 kph) in a matter of seconds? Cars can be fueled with natural gas, which generates less carbon dioxide than gasoline does.

Fuel cells can also power cars, using water and electrodes that generate a current, but they need a ready supply of hydrogen. At the moment, however, this technology is still some way off. (Interestingly, one can drink the exhaust of a car powered by hydrogen, because when hydrogen burns it produces pure water.)

Unfortunately, major US auto companies have resisted and continue to resist thinking creatively—a stance that has caused the devastation of Detroit and hundreds of thousands of working-class families. Years ago, these companies knew how to make efficient automobiles driving at 50–60 mpg, but they stubbornly stuck to old, outdated designs. But now that the writing is on the wall, US car manufacturers have even resorted to copying the latest Japanese designs—a rather sorry setback for an industry that once led the world in automobile technology. The US auto industry could have overtaken the Japanese industry by manufacturing solar-, hydrogen-, and fuel-cell cars years ago, while at the same time helping to save the planet.

Ethanol has become the new "savior" that will simultaneously reduce global warming and function as a renewable resource. The facts, however, do not validate these assumptions.

ETHANOL: FOOD FOR CARS, NOT PEOPLE

By investing heavily in ethanol as an energy source, Brazil has become somewhat energy-independent. In 1988, sugar-based ethanol production provided 62 percent of Brazil's automotive fuel. Recently, as the oil crisis has become more severe (particularly in the United States) and the threat of global warming has become undeniable, politicians, farmers, and the public decided to emulate Brazil by undertaking a sudden conversion to cars fueled by alcohol (ethanol). Although huge amounts of money have been allocated by Congress for this new industry and alcohol gives off 63 percent less carbon than does gasoline, enormous problems are associated with ethanol as a fuel.

The explosion in farm-grown biofuels has elevated grain prices to record highs and induced both a massive deforestation of the Amazon jungle and an inflation of global food prices that is leaving a hundred million people hungry. US farmers previously grew a high percentage of the world's soybeans, but now they use large areas of land to grow corn for alcohol—which accounts for one-fifth of their corn crop. In Brazil, soy for livestock feed is now grown on cattle pastures while more and more forest is cleared for cattle grazing. The grain that is used to fill a single SUV tank just one time could feed one person for a whole year. Former UN rapporteur on food, Jean Ziegler, called biofuels "a crime against humanity."[32] Lester Brown from the Earth Policy Institute says that biofuels pit eight hundred million people with cars against eight hundred million people who are hungry.[33]

The United States, confident that biofuels will help alleviate global warming, has quintupled biofuel production over the last ten years, and Congress just mandated another fivefold increase over the next decade. Europe is exhibiting a similar zeal for alcohol-powered cars. Global investment in biofuels increased from $5 billion in 1995 to $38 billion in 2005, and it is predicted to be $100 billion by 2010, aided and abetted by Richard Branson of Virgin Airlines, George Soros, BP, GE, Shell, Ford, Cargill, and the Carlyle Group.[34]

However, these people are cutting off their nose to spite their face, for alcohol derived from grain is absolutely disastrous for global warming from a scientific perspective. Using land to grow fuel is destroying forests, wetlands, and grasslands, which store huge amounts of carbon. Deforestation accounts for 20 percent of global warming. Malaysia is destroying her forests to grow palm oil for biodiesel, and Indonesia has razed so much wilderness that it is now one of the world's leading carbon emitters. All this destruction is backed by billions of dollars in investment capital.[35]

An area the size of Rhode Island in the Amazon was deforested in the last half of 2007. As trees are lost, the forest dries out, rainfall decreases, and fires burn large swaths of forest. The rate of deforestation is closely correlated with the commodity prices on the Chicago Board of Trade. A recent study in *Science* estimated that the subsequent carbon emissions from deforestation mean that corn and soy biodiesel actually produce double the emissions of a similar amount of gasoline.[36]

If used the right way, however, biofuels can have a productive future.

- They can be made from a variety of substances, including crop wastes, wood chips, scrap plastic, rubber, and municipal garbage.[37]

- As demonstrated by a recent, exciting development, they can be made from seaweed, which can be cultivated and harvested using no soil and no extra water.[38]
- The production of algae, which grow very fast, can use CO_2 produced from the burning of fossil fuels as a fertilizer.[39]

Biofuels have become so popular in the United States that Iowa, the corn capital of the world, now has fifty-three thousand jobs and $1.8 billion of its economy dependent on biofuels. Iowa has surprisingly become a net importer of corn, while the ethanol boom has helped lift farm incomes to record highs nationwide. But a study by David Tilman, a University of Minnesota ecologist, shows that it will take over four hundred years of biodiesel use to "pay back" the carbon directly emitted by clearing peat bogs to grow palm oil, and ninety-three years to use up the carbon from clearing grasslands to grow corn for ethanol. Biofuels increase the demand for food crops while boosting prices; the rise in prices then drives further agricultural expansion, decimating forests. Overall, if all these ramifications are taken into account, corn biofuel has a carbon payback period of about 167 years![40]

Fact Number Three: Bicycles Are the Key to Global and Human Health

Bicycles use human energy and save global energy. They are clean, efficient, and healthful for human bodies. Roads must give way to bicycle tracks. When I visited China in 1987, special avenues teaming with bicycles of five or six lanes were separated from roads used for motorized traffic and pedestrians. In fact, I did not see one car while I was in China for a period of six days. As mentioned earlier, however, bicycles are now tragically forbidden on many streets in big cities in China, and more and more roads are being constructed to accommodate cars.

It is now imperative that a large percentage of the population of the United States and the Western world emulate the old China and take up bicycle riding, which not only will help prevent global warming but is excellent for the musculoskeletal and cardiovascular systems. Safe bicycle lanes must be built in all cities and suburbs on an urgent basis. Distant, large-scale supermarkets and shopping malls accessible only by car will become obsolete as people demand small, convenient shops within walking or riding distance of their homes. Society must reestablish small community shopping centers, where people meet each other and socialize, and where the emphasis is on the community rather than on consumerism. What a healthy, exciting prospect!

Fact Number Four: Energy Efficiency and Conservation Are an Imperative Response to Global Warming

Buildings can be made extremely energy-efficient. Improved designs for stoves, refrigerators, and electric hot-water heaters increase energy efficiency by between 5 and 87 percent. The sealing of air leaks in houses can cut annual fuel bills by 30 percent, and double-paned, insulated windows greatly reduce energy loss. Superinsulated houses can be heated for one-tenth the average cost of heating a conventional home. In the United States, 20 percent of the electricity generated is used for lighting, but new compact fluorescent bulbs are 75 percent more efficient than conventional bulbs. In 2001, US households accounted for 8.8 percent of the country's total electricity consumption.[41] Theoretically, then, the country could reduce its electricity use for lighting to 5 percent of the total. This, together with other simple household conservation measures, would enable the closing down and mothballing of all nuclear reactors in the United States, which currently generate 20 percent of all the electricity used.[42]

I lived in the beautiful city of Boston for fourteen years and grew to love the old New England houses. But now that I am more aware of the fate of the Earth, I realize that these are totally inappropriate dwellings. They are big, rambling, leaky, and inefficient. The vast quantities of oil required to heat these large volumes of enclosed air through a long Boston winter add substantially to carbon dioxide–induced global warming. When these handsome houses were built, in the nineteenth and twentieth centuries no one imagined that the Earth would someday be in jeopardy. Fuel supplies seemed endless, and the air was relatively clean.

Houses of that kind can be made somewhat fuel-efficient by the insulation of walls, ceilings, and windows and by the sealing of all leaks. To encourage such reform, the federal government must legislate adequate tax incentives, for in the long run these will provide insurance for our children's future.

In contrast, solar buildings are now in an advanced stage of design and development. The need is for legislation that requires all new buildings, residential and office, to be solar-powered, with large, heat-trapping windows oriented toward the south; floors made of tile and cement, which trap the sun's heat during the day; and appropriate window insulation, which retains the heat at night. Solar hot-water panels and solar electricity generation are relatively cheap and state-of-the-art. Firms that manufacture such equipment are now making large profits. Home owners would benefit because they would become independent of utilities; they could even sell back electricity to the local utility at off-peak hours. Indeed, increasing numbers of Americans are already vendors of electricity. Some state governments are taking a lead in this new field. In 2007, California Governor Arnold Schwarzenegger mandated that one million solar houses be built in California by the year 2020.

Solar technology will inevitably become highly efficient and cheap, and a huge market will open up in the developing world. This has already happened in China. Industrialized countries could assist billions of people in bypassing the fossil fuel era, helping them to generate electricity from solar and wind power and to use solar cookers and solar hot-water generators—a signal solution to the problem of ongoing global warming. The developed world must help the developing world bypass the fossil fuel era if the Earth is to survive.

Attention should also be given to the strange high-rise buildings covered in tinted glass that seem to be in vogue in many US cities. These are not solar buildings. The windows cannot be opened to allow ventilation during the summer, so they must be cooled with air conditioners, which use ozone-destroying hydrochlorofluorocarbons and carbon dioxide–producing or nuclear waste–

producing electricity. In the winter, heat leaks from the windows like water through a sieve. And these buildings are generally lit up like Christmas trees at night, for no apparent purpose, by energy-inefficient lighting. Dallas, Houston, and Los Angeles are just a few of the many US cities that boast numerous of these monstrosities.

We all must become acutely conscious of the way we live. Every time we turn on a switch to light a room, power a hair dryer, or toast a piece of bread, we are adding to global warming. We should never have more than one lightbulb burning at night in our house, unless there are two people in the house in different rooms. Lights must not be left on overnight in houses or gardens for show, and all lights must be extinguished in office buildings at night.

Note also that motorized lawn mowers produce forty times more emissions than does a car, and one hour's use of an outboard motorboat emits the same amount of pollution as do fifty cars operating at a similar speed. These particular boat engines are banned in the United States because of more stringent emission regulations than in other countries. Small engines on leaf blowers and weed whackers are similarly polluting.[43]

Clothes dryers are ubiquitous and unnecessary, and in the United States they account for 6–10 percent of all residential electricity use, equivalent to the output of fifteen of the nation's nuclear reactors. In Australia, we dry our clothes outside in the sun, hung by pegs from a line. Americans can do the same in the summer, and in winter in the colder climes, like Boston's, clothes can be hung on lines in the cellar near the furnace. In some American cities, laws prohibit people from hanging clothes on lines outside, on grounds that it's not aesthetically pleasing. In fact some sixty million Americans live in "association-governed" residential communities that restrict clotheslines because they are seen as visual pollution or as a sign of poverty, which is ridiculous.[44] This solar method of drying offers, in fact, an easy and efficient step toward the reduction of atmospheric carbon dioxide and radioactive waste. And bear in mind that electrically operated doors, hand dryers, escalators, and elevators all contribute to global warming. Although it is necessary for people with disabilities to use elevators and escalators, able-bodied people should use their legs to save electricity.

At a most sensitive time in the history of the planet, when global warming threatens one-third of the Earth's species and large sections of the human race, and there is an urgent mandate to cease burning fossil fuels, very serious news emerged that big oil companies have regained control of the third largest oil deposits in the world—in Iraq. On June 19, 2008, the *New York Times* reported that some four Western oil companies had negotiated no-bid contracts that would allow them back into Iraq after being expelled in the 1960s by Saddam Hussein when he nationalized Iraqi oil, transferring it from foreign oil companies to the Iraqi people. In fact, there are five companies: US-based ExxonMobil and Chevron, Royal Dutch Shell, France's Total, and Britain's BP.[45] On September 11, 2008, however, it was reported, again in the *New York Times*, that the deals had been negated when, following sharp criticism from several US senators, Iraqi Oil Minister Hussain al-Shahristani had withdrawn the agreement.[46] Subsequently, on September 22, the government of Iraq and Royal Dutch Shell signed a deal allowing Shell access to Iraqi natural gas.[47]

Only belatedly and intermittently have the major US media discussed the fact that the invasion of Iraq by George W. Bush's administration was about oil. Rather, as administration sycophants, the media have continued to repeat, ad nauseam, the line that there were "weapons of mass destruction" in Iraq, and that Iraq had been involved in the 9/11 attack on the World Trade Center, and therefore Saddam Hussein needed to be eradicated.

Now we know the truth. One of the few buildings that was not attacked and decimated during the shock-and-awe attack on Baghdad in 2001 was the Oil Ministry. Since that time it has become transparent that the invasion—during which, according to some estimates, over one million Iraqi civilians have been killed by "weapons of mass destruction" including cluster bombs, daisy cutters, fuel air explosives, depleted uranium weapons and many others,[48] and more than three million Iraqis have become refugees—was all so that the United States could regain control of the Iraqi oil deposits.

Debate on Global Warming

Now let's look at the way the US public debate on global warming has been orchestrated. Although some oil companies, including BP, Occidental Petroleum, and Shell, recently responded to calls to reduce their carbon emissions and to invest in clean energy technologies, ExxonMobil has made no such commitments.

ExxonMobil, the world's most profitable company in history, is the largest publicly traded company. In 2005, its revenues, at $339 billion, exceeded the gross domestic product of most of the world's nations, representing $100 million a day in profit.[49] In 2008, the ExxonMobil Corporation made $11.68 billion in profits in the second quarter, the highest income ever recorded by a US company.[50] In its manufacture of end products that include kerosene, aviation fuels, gasoline, heating oil, diesel products, and heavy fuels, it is also one of the world's largest produces of greenhouse gases.

In the thirteen years leading up to 2006, the company was led by Lee Raymond, who stringently opposed curbs on CO_2 emissions and steadfastly refused to acknowledge the scientific data on global warming. Paul Krugman from the *New York Times* called Raymond "an enemy of the planet."[51] For his efforts, Raymond was paid $686 million while at the helm of ExxonMobil, and he received a $400 million bonus on retirement.[52]

ExxonMobil has been the most active corporation funding efforts to negate climate change regulation. It distributed $16 million between 1998 and 2005 to a network of disadvocacy organizations such as the American Enterprise Institute, the Cato Institute, the Competitive Enterprise Institute, and many other lesser-known organizations to manufacture public and political "uncertainty" on the scientific evidence of global warming. Many of these organizations have overlapping, sometimes identical groups of people acting as spokespersons on the issue of global warming, thus inducing a magnifying effect making it appear that information is coming from many sources.[53]

An excellent investigative report on ExxonMobil by the Union of Concerned Scientists documents how the tactics and strategies used by this oil company were and are the same as the campaign that the tobacco companies employed to prevent government regulation of their carcinogenic industry—that is, to create public confusion about the link between smoking and disease through the manufacture of scientific uncertainty in the minds of the public and the politicians.[54]

ExxonMobil also was heavily involved in political funding and lobbying, paying lobbyists $61 million between 1998 and 2005 to gain access to key decision makers, funding the Bush reelection campaign, and giving $935,000 in political donations for the 2004 election.[55] Furthermore, not only are the oil companies making obscene profits, but they have managed to persuade Congress to allocate to them $33 billion in tax-funded giveaways over the next five years.[56]

Of note, too, is the fact that Vice President Dick Cheney, who also supported these activities of ExxonMobil, has been instrumental in shaping the energy policy of the United States over the past eight years.[57] On April 17, 2001, soon after George W. Bush was first elected, Cheney met with Kenneth Lay (now deceased) from the Enron Corporation to discuss "energy policy matters." Lay gave Cheney a three-page wish list of corporate recommendations, of which seventeen were adopted by the National Energy Policy Development Group in seven of eight policy areas.[58]

Cheney met with Lay at least six times while drawing up the US energy policy, as well as with people from the coal, gas, oil, and nuclear companies, all of whom had contributed significant funds to the Bush campaign. Cheney steadfastly refused to release the documents related to those meetings, even though the Federal Advisory Committee Act of 1972 dictates that task forces such as Cheney's must conduct public meetings and that publicly available records must be kept.[59] No wonder there has been little government movement on the installation of renewable energy systems throughout the United States over the last eight years!

For many years, the details of the business of energy production have been jealously guarded by the utilities, oil and nuclear companies, and government departments, which often work together secretly.[60] As the *New York Times* reported on November 6, 1989, "The Energy Department's reliance on contractors and consultants for basic Government functions is pervasive but largely outside public scrutiny . . . The department has increasingly retained consultants to help perform 'virtually all basic functions.'"[61] The situation was so bad that in 1989 the secretary of energy, Admiral James D. Watkins, was extremely embarrassed to discover that testimony he had given to Congress had actually been written by the very corporations that are supposed to be regulated by the Department of Energy (DOE).[62]

Because the energy policies of the DOE are determined by the vested interests of its contractors, almost all of the emphasis has been placed on coal, oil, and nuclear power, but virtually none on alternative energy techniques—a situation reflected by the fact that only 2 percent of US electricity is generated from renewable sources.[63] President Obama may change this.

The US Environmental Protection Agency was also seriously compromised under the Bush/Cheney administration. The vice president of the EPA's scientists' union, J. William Hirzy, charged that the Bush administration's tactics amount to "political interference": trying to starve the EPA of cash, including allocating record-low funding for community drinking-water facilities and for the Superfund hazardous-waste cleanup program, and using scientists to justify the political goals of the White House, which are to support the interests of big business. The Bush/Cheney White House even directed the EPA to estimate the cost of proposed regulations to big business—industry was calling the shots. The agency, under the direction of Administrator Stephen Johnson, summarily closed some of its libraries around the United States, eased requirements that direct companies to provide data about the volume of hazardous chemicals they release into the air and water in a given community, and limited public access to reams of federal information on pollutants, industries, and agency research. The EPA also reversed its stand on the Californian initiative to limit tailpipe emissions from cars and trucks, thereby blocking similar initiatives in more than a dozen other states that wanted to follow suit. The decision, reversed by President Obama in January 2008,[64] was applauded by the auto industry and, of course, supported by the White House, which has opposed mandatory caps on emissions.[65]

In October 2007, Dick Cheney's office edited out six pages of the congressional testimony of Julie Gerberding, director of the Centers for Disease Control and Prevention, which detailed some of the health threats posed by global warming.[66] In other words, the Bush administration protected corporations at the expense of its citizens' health. This, if I may say as a physician, is criminal behavior. I wonder if Dick Cheney has ever seen or helped a child die of cancer or tried to comfort the parents in their grief, never to recover? In truth, every child's life is as precious as the life of Cheney's own grandchild.[67]

Now is the time for the politicians and the energy companies to seriously commit themselves to addressing the pressing problem of global warming. No more shilly-shallying with one-upping the public, no more lobbying and obfuscation of data by wealthy corporations, no more games played by politicians to please their corporate donors and special-interest groups. This is not business as usual. It is an emergency, and we cannot play Russian roulette with the fate of the Earth. Interestingly, society is always ready in time of war to drop everything and mobilize to save the nation-state. During the Second World War, for instance, following the Japanese bombing of Pearl Harbor in 1942, the US industrial economy shifted to producing planes, tanks, and guns within months, and a ban on the sale of cars lasted three years.[68] This is the sort of psychological, pragmatic, and urgent response that we now must make to the threat of global warming. The eminent climatologist James Hansen said in December 2007, "We as a species passed climate tipping points for major ice sheet and species loss when we exceeded 300–350 ppm CO_2 in the atmosphere, a point passed decades ago . . . We either begin to roll back not only the emissions of CO_2 but also the absolute amount in the atmosphere, or else we're going to get big impacts . . . We should set a target of CO_2 that's low enough to avoid the point of no return (300 to 350 ppm). We have to figure out how to live without fossil fuels some day . . . Why not sooner?"[69]

To reach these objectives, James Hansen makes four broad suggestions:[70]

1. Phase out existing coal-fired plants over the next ten to twenty years, and build new ones only if adequate carbon capture and storage (CCS) is developed, which, as discussed earlier, is ten to twenty years from pragmatic implementation. (Without CCS, a substantial fraction of the CO_2 released from new plants would stay in the air for over five hundred years.)
2. Severely ration oil and gas, using it only for transportation until the industrial era moves "beyond petroleum." Apply a universal carbon tax that is sufficient to severely discourage the energy-wasteful and dangerous extraction of tar shale oils and liquid fuel from coal. The era of hybrid and electric cars is upon us, taking us out of the fossil fuel–powered transportation era.
3. Immediately focus great attention on non-CO_2 greenhouse emissions, including methane, nitrous oxides, soot, tropospheric ozone, and CFC gases.
4. Increase the absorption of greenhouse gases from the atmosphere by encouraging the growth of forests, undertake farming practices that encourage carbon retention and storage in the soil, and protect ocean absorption systems. Replace grain-based biofuels with biofuels from possible other sources, such as cellulosic native grasses, hemp, kenaf, and algae.

What is to be done about the fossil fuel problem that Hansen enumerates in his first two points? That's easy. A hard cap must be imposed on the big emitters of carbon by 2010—in 2005, the largest emitters produced 70 percent of the fossil fuel–derived carbon. These industries would purchase auctioned allowances from the EPA, the cost of which would depend on market value, and the total amount of allowances permitted would decrease each year until it reached zero between 2030 and 2050.[71]

James Hansen recommends that this money be returned to the

people so that they can use it to insulate their houses and provide their own solar or wind power. Personally, I think the money should be used by the federal government to produce solar energy in a variety of forms, to convert the national auto fleet to a combination of hybrid and all-electric cars, to provide efficient and ubiquitous mass-transit systems, to create massive wind farms, to upgrade the national grid, to install geothermal power plants in appropriate places, and to implement tidal and wind power installations. Appropriate subsidies should be given to home owners to install solar systems and to make their houses more energy-efficient. Similarly, industry will need help installing cogeneration systems and radically changing technologies to become more energy-efficient.

Such reforms will generate hundreds of thousands if not millions of jobs, boost the GDP, and help make the United States the leading energy nation of the world.

4 Trees: The Lungs of the Earth

A tree is a noble organism, unique in its beauty and a home and refuge for birds, insects, and small animals. The eminent biologist E. O. Wilson found on one tree in Peru forty-three ant species, belonging to twenty-six genera—approximately equivalent to the diversity of ants in the United Kingdom.[1]

I live in Australia, where gum (eucalyptus) trees predominate. There are more than six hundred varieties, and even though many are similar in appearance, each tree has a design different from that of any other. When I was growing up, I thought gum trees were boring and all the same. It was not until I had lived in the Northern Hemisphere for fourteen years that I began to really appreciate the unmatched flora and fauna of Australia.

When the English invaded Australia some two hundred years ago, they destroyed large areas of native bush in order to plant rosebushes and deciduous Northern Hemisphere trees and to create rolling meadows reminiscent of the English countryside. They clearly felt uncomfortable with odd-shaped animals like kangaroos, screeching colorful parrots, and strange prickly bushes and trees. Originally, 14 percent of this vast desert continent was forested;

now only 7 percent of that forest remains. Since Captain Cook explored Australia in the eighteenth century, forty-one bird and mammal species and over a hundred plant species have become extinct. The list of endangered species includes 10 species of fish, 12 frogs, 13 reptiles, 32 birds, 33 mammals, and 209 plants. Many more species are listed as "vulnerable" or "rare."[2] We have cleared 75 percent of our rain forests since the late 1700s,[3] and Australia has more than three times as much degraded land per capita as do comparable countries.[4]

The trees were felled and the land cleared for agriculture. But over the last thirty-five years, the government has organized and encouraged aggressive deforestation and land degradation by cutting down our last remaining forests, and sending the tree corpses to a Japanese-owned chip mill in New South Wales. The wood chips are then shipped to Japan to make paper that is used in glossy magazines such as *Vogue*. So alarming is the rate of deforestation that, if present trends continue, the koala may become extinct as its habitat contracts and disappears. In 1990, I was told about a new development adjacent to a main highway where a real-estate agent had decided to fell a stand of magnificent gum trees. The following day, koalas and echidnas were staggering across the freeway and being hit by speeding cars.

Recall that James Hansen said that one of the ways to alleviate global warming and to absorb anthropogenic CO_2 is through forests.[5] The truth is that up to 30 percent of Australian CO_2 emissions emanate from the destruction and desecration of the remaining Australian forests. This estimate includes CO_2 emissions from damaged soil, the burning of debris from the trees after logging, and the production of paper from the wood chips. Surprisingly, the forestry industry has such a hold on the Australian government that the CO_2 that will eventually be released from massive quantities of exported forest logs is simply not accounted for. Nor is the CO_2 emanating from global logging factored into the Kyoto carbon account balance or the Stern report,[6] on the grounds that these native forests will regrow. This is fatuous ideology because much of the cleared land will be used for farming. Besides, it takes up to five hundred years for a tree to recapture the CO_2 released after the felling of an old-growth tree.[7] Once upon a time, the forests in the United States stretched from coast to coast, now they are almost all gone, except for some stands on the West Coast that are continually being logged by paper companies such as Weyerhaeuser.

In terms of the biology of the planet, "development" is a euphemism for destruction. Even the frequent use of the term "sustainable development" is an exercise in obfuscation. In a world where all resources are finite—forests, minerals, soil, air, and water—continued use and abuse of them can have only one end: the depletion and destruction of most life. Once a forest that has taken thousands of years to evolve into a system of complex biodiversity is destroyed, it takes hundreds of years to regenerate.

All trees are precious. They are more than just havens for animals, birds, insects, and humans; they are also the lungs of the Earth. Just as we breathe oxygen into our lungs and exhale carbon dioxide, so trees breathe carbon dioxide into their leaves and exhale oxygen. Trees are really upside-down lungs: their trunks are equivalent to the trachea, their branches to the right and left main bronchi, and all their branching twigs and leaves to small bronchi and alveoli, or air sacs, where oxygen and carbon dioxide are exchanged. Tree trunks and branches may appear solid, but they are really rigid channels that transmit water and nutrients to the leaves, the way the trachea and air passages transmit air to the alveoli in human lungs.

Trees are therefore an organic necessity to the biological health of the planet. As human beings fill the air with carbon dioxide and destroy the ozone layer with human-made chemicals, trees offer an excellent means of buffering these effects. It has been calculated that if an area the size of Australia or the United States were planted with trees, the air could be cleared of carbon dioxide released from fossil fuels.[8]

Just as the planting of trees replenishes the atmosphere, so deforestation helps destroy it. Because a tree spends two hundred to five hundred years absorbing carbon dioxide and storing the carbon in its wood, when we chop it down and burn it as either wood or paper, we release up to five hundred years of trapped carbon as carbon dioxide, thus exacerbating the greenhouse effect. The ozone layer and greenhouse gases do not recognize national boundaries, so every tree felled has global ramifications. We need to stop felling trees and plant many more of them.

Once upon a time in the millennium before Christ, the countries of the Middle East were covered with a humid tropical rain forest. Over time, the trees were chopped down for boats and the development of civilization. Now these countries are virtual deserts. In fact, most countries in the Northern Hemisphere were covered with forests that teemed with life; now the trees and the wildlife are almost gone. According to the United Nations, about 40 percent of Central America's forests were destroyed between 1950 and 1980, and during the same period Africa lost about 23 percent of its forests. When forests vanish, the climate tends to change, making natural reforestation almost impossible. In some arid climates, like Israel's, intensive drip irrigation has been used to initiate reforestation programs. But it will be many years before these trees reach such a mass that they will affect the climate.

In November 1989, I went to Puerto Ayacucho, in Venezuela, hired a dugout canoe, and set off down the Orinoco River, a tributary of the great Amazon River. All my life I had dreamed of the Amazon forest, and as I read of its impending destruction, I knew I had to see and experience it while it still stood in its magnificence. I took my twenty-four-year-old son, William. Our boat, manned by a driver, a cook, and a guide, was at least fifty feet long, made from a single tree. We slept in hammocks located on the prow and spent ten days on the river. The climate was hot and humid, almost unbearable when we pulled in at huge sandstone rocks jutting out into the river to cook our meals, but simply beautiful as we moved slowly along the river and the soft perfumed breeze slid past our faces. I have never smelled such pure, scented air anywhere else in all my life.

We would wake up early and embark on the day's journey at sunrise. William and I lay rocking in our hammocks, gliding into the rosy pink sky surrounded by the most magnificent forest, filled with exotic trees I had never seen before. We encountered hundreds of different palms; huge flowering trees covered with pink, white, or yellow blossoms; and colorful birds screaming as they flew from the edge of the river deep into the jungle. The undergrowth was so thick and matted as to be impenetrable. At night, white freshwater dolphins snorted and spurted in the water, while the jungle was alive with raucous screams of monkeys and other animals. As we lay under our mosquito nets, the sky was ablaze with the brightest stars I had ever seen.

When we stopped the boat, we were immediately covered with biting insects of all sizes and shapes. The bees were ten times larger than any I had observed before, and tiny midges, though hardly visible, left a subcutaneous hemorrhage and an intense itch that lasted six weeks. It took great willpower not to scratch, and despite the application of the most carcinogenic insect repellent used in Vietnam, the insects were not deterred. I decided that jungles are not for people, but for insects.

The jungle went on forever—an ocean of trees, covering an area the size of the United States. It seemed to me that humans could never have a significant impact on this vast creation of nature. But then I remembered the Middle East and the Sahara desert and realized that, in those days, it took two men nearly a week to fell a single tree. Now the chainsaw, bulldozer, and matches are much more efficient.

Every day in the Amazon, each tree transpires into the air hundreds of gallons of water, which evaporates and creates beautiful, billowy cumulus clouds that tower above the river and unleash a deluge lasting one to two hours. When the forest is destroyed, the transpiration of water ceases, the rain stops, the soil dries out, and the region becomes a desert. It is said that if half the Amazon jungle is cleared, the existence of the remaining forest will be threatened by decreased rainfall.[9] In addition, the Amazon river system is the world's largest. So great is its outflow of water that fifty miles out to sea you can still drink fresh water from the ocean. The river is surrounded by half the world's rain forest, which is home to almost 10 percent of the world's mammals, and 15 percent of the world's known land-based species.[10]

The Amazon seems to me a perfect example of the microclimates created by forests, and of the change of climate that ensues when humans destroy the cathedrals of nature. As my son and I started our journey, we noticed clearings along the river populated by primitive buildings adorned with crosses. These were Indian villages. We stopped at one of them, and the chief came out to greet us, so drunk that he could hardly stand. I realized with a sense of shock that these Yanomami Indians had been "civilized" by Catholic missionaries and taught to adopt a Western lifestyle. They wore skirts, or T-shirts and trousers, and seemed uncomfortable in them. In the past, many had been infected with our diseases, such as malaria, venereal disease, tuberculosis, influenza, and measles, and had died. Others had become addicted to our drugs, and all had lost their innate ability to live with and from the forest, eating the food the forest provided; they now existed, to a large degree, on Western diets of flour, sugar, canned foods, and so on.

As we traveled deeper into the forest, these villages became less common, until eventually we met a canoe of naked men sitting proud and tall in their boat, armed with spears and bows and arrows, painted with ocher, and adorned with feathers. They were off for a day's hunting. A little farther down the river, we came upon a group of women and children at the edge of the water. There was no clearing, but we saw tenuous tracks through the matted jungle. Some of the younger children had swollen bellies; the older children and women, by contrast, were not only healthy but were the most curious, alive, and vibrant people I had ever seen. They climbed into our boat, and one of the women pointed to my brassiere, which was drying on the side of the canoe, indicating that she would like to try it on. I reluctantly agreed, and within five minutes, twenty women had squeezed themselves, one after another, into this garment before handing it on to others. It turned from white to dark brown, and as we departed I decided, with some misgivings, to leave it with them, since they liked it so much. One more token of Western civilization to pollute their way of life!

The more I saw of the indigenous people, the less I thought our culture was at all civilized. They live in harmony and peace with the forest, protecting and respecting it, while we rape and destroy it for economic reasons.

The Indians have for thousands of years practiced a form of cultivation in the jungle that is ecologically sound. They fell a group of trees and allow them to fall outward radially. They then burn the leaves on the outer circle to fertilize the soil and sow plants and fruits in a radial pattern, the plants that need more fertilizer being in the outer circle. Eventually, the forest encroaches on the garden, and within ten to fifteen years the cultivated circle reverts back to forest.[11]

Indians have coexisted with the Amazon for ten thousand years. But now the Yanomami Indians are being massacred by local governments, by miners, and by the army because they are regarded as an obstacle to "development." Brazilian laws recognize Indians as wards of the state with rights similar to those of minor children. When the Portuguese invaded Brazil in the 1500s, there were five million Indians. Now there are approximately two hundred thousand from 180 different indigenous nations.[12] More than ninety different Amazonian tribes are thought to have vanished in the twentieth century.[13]

When we finally entered Brazil from Venezuela and Colombia after our ten-day voyage, we began to see the ravages of civilization. The Brazilian government has appropriated large tracts of jungle to relocate poor peasants from the cities. The forest has been destroyed, and the dead giants are left rotting on the ground as the peasants attempt to make a living by planting spindly banana plants and other trees among the forest debris. The trouble is that the soil of the Amazon is relatively sterile, and once the forest disappears, the humus composed of ever-recycling dead leaves and bacteria is eliminated. Because there is no ongoing regenerative composting, the soil becomes unsuitable for agriculture. The peasants are struggling to make a living, and the forest is dying.

The World Bank and the International Monetary Fund, in their misguided beneficence, built a road called BR364 right through the middle of the jungle for "development." Side roads have subsequently been constructed, and the whole forest has been opened up for destruction.[14]

The Amazon is also threatened by several other major destructive enterprises. But it is important to bear in mind the following facts as you review the various methods of rain forest destruction occurring today:

- The global tropical rain forests, which are the oldest continuous ecosystems in the world, are being destroyed at the rate of one football field per second, or an area twice the size of Rhode Island per week.
- Tropical rain forests boast on the order of a thousand species per square kilometer, compared with only about hundred species in a similar area in North America.
- An area of tropical rain forest the size of Florida is cleared each year.[15]
- Tropical rain forests contain about 50–80 percent of the world's species of plants and animals, of which there are estimated to be thirty million.[16]
- At the present frantic rate, deforestation will destroy all the world's tropical forests within twenty-five to fifty years, along with fifteen million to twenty-four million species, thereby turning the land into a desert.
- The Amazon jungle is thought to house sixteen hundred bird species, and just two and a half acres of jungle may contain forty thousand different species of insects.
- Rain forests are the oldest living ecological systems on Earth, having evolved over millions of years.[17]
- The Club of Earth maintains that species extinction is "a threat to civilization second only to the threat of nuclear war.[18]

Clearing and Soy Production

This story should make people's hair turn gray. According to a Greenpeace report, three US-based agricultural giants—Archer Daniels Midland (ADM), Bunge, and Cargill—control more than three-fourths of the soy-crushing facilities in Europe. Since January 2003, almost 27,000 square miles (about 70,000 square kilometers) of the Amazon rain forest in Brazil has been cleared—at the rate of one football field every eight seconds. Slave labor plays a big part in this rain forest destruction, and about 75 percent of Brazil's greenhouse gas emissions come from clearing and burning of the forest. The soy is sold and used for animal feed in Europe, which buys half the soy grown in the denuded Amazon forest. Eighty percent of the world's soy production is fed to the livestock industry.[19] So the Amazon is being chopped down for the profits of these US corporate giants to produce food for meat. This dynamic is driven partly by the diversion of US corn—traditionally a food source—to ethanol production for cars.

Fires

At the end of the dry season, the burning season begins in Brazil across the Amazon region, pushed by illegal loggers, cattle ranchers, soy producers, and wildcat miners. In October 2007, government satellites counted a total of 16,592 illegal forest fires burning across Brazil. According to one of the pilots monitoring the situation from the air, the scene looked like *Apocalypse Now*.[20]

According to Conservation International, the Amazon is at great risk of unprecedented damage from plans to improve transportation, communications, and power generation in the region. Improved transportation networks will make it easier for inaccessible areas to be logged and burned, disrupting precious ecosystems that support native species and threatening indigenous populations.[21]

Lumbering

Brazil owes a huge debt to US banks. The World Bank and the International Monetary Fund, which represent the interests of these banks, have advised the Brazilian government to chop down the Amazon forest to pay back the debt. In the past, they have also funded Brazil's military dictators to build dams and encourage the resettlement of thousands of people from the destroyed jungle.[22] Commercial logging is the second most serious cause of deforestation. Slash-and-burn destruction for agriculture ranks number one. In a single day in 1988, six thousand separate fires were burning in the Amazon forest—all human-made.[23] In 1980, it was estimated that the burning of tropical forests added an extra 1.7 billion tons of carbon dioxide to the air every year.[24] Logging is difficult in dense jungle, and often twenty trees are destroyed to obtain one tree suitable for timber. One-third of the felled trees are destroyed to obtain access for logging equipment. The harvested timber is used for charcoal and firewood, for paper production, and for the building industry in developed countries.[25]

Japan imports over 40 percent of the world's tropical timber. It consumes the equivalent of one forest per day in disposable wooden chopsticks, or some twenty billion per year.[26] Much beautiful tropical wood is used in Japan as disposable plywood molding in the construction business.[27] The United States imports $2 billion worth of tropical wood per year, and China has become the leading importer of tropical wood. In 2006, China exported $8.8 billion worth of wooden furniture, $4.2 billion of which went to the United States.[28] Because of the high demand for such wood, tropical countries are exhausting their reserves, as well as their potential earnings, from renewable forest products such as Brazil nuts. Most rain forest products need a fully functioning rain forest system to produce efficiently. Developing countries are therefore jeopardizing their use of forest resources through deforestation.[29]

Mining

The soil beneath the Amazon forest houses valuable gold deposits, and half a million people are, at present, digging up and destroying the forest in their quest to find this treasure. Brazil is now the fifth largest gold producer in the world; in 1988, more than 182 tons of gold was extracted from the jungle floor. Mercury is used for the gold extraction process, and this toxic heavy metal is then released into the river. Indigenous populations living downstream rely on this polluted water for drinking and washing. Mercury poisoning is very serious; it causes kidney damage and Minimata disease, as well as brain damage and mental retardation in babies and children. The forest also contains the richest iron ore deposits in the world. Once extracted, the iron is smelted in fires fueled by charcoal, made from partly burned trees. By 1992, seven smelting factories were consuming seven hundred thousand tons of charcoal per year.[30]

Oil and Gas Projects

A study by two US nonprofit organizations (Save America's Forests and Land Is Life) and scientists from Duke University published in August 2008 revealed that over 180 oil and gas "blocks," or areas zoned for exploration and development, cover 688,000 square kilometers of the preciously diverse area of the western Amazon in Bolivia, Colombia, Ecuador, Peru, and western Brazil. These blocks perfectly overlap the most biodiverse part of the Amazon for birds, mammals, and amphibians and of territories of indigenous people in voluntary isolation. Driven by growing global energy demand for ever-more oil and gas, companies from the United States, Canada, Europe, and China lead the push. This ever-insatiable desire for fossil fuel, coupled with corporate greed, will wreck some of the last remaining bastions of wilderness and sanctuaries of endangered species and people.[31]

Dams

The World Bank recently planned the construction of 125 hydroelectric dams in the Amazonian jungle by the year 2010, which would flood six hundred thousand acres (nearly 900 square miles, or 2,300 square kilometers) of rain forest.[32]

Cattle Ranching

Ranching occupies 72 percent of the cleared forest areas, and most of the beef is used to supply fast-food burger chains in the United States, Central America, and Europe. Between 2000 and 2005, 60–70 percent of Brazil's forest area was cleared for cattle ranching, making cattle ranching the leading cause of deforestation in the Brazilian Amazon. According to the Center for International Forestry Research (CIFOR), the percentage of processed meat imported from Brazil to Europe rose from 40 to 74 percent between 1990 and 2001. In 2003, 80 percent of the cattle produced in the Brazilian Amazon were exported.[33]

Amazon cattle are cheap to raise, but the fertility of the cleared land is diminished. Good grass grows for only two years before the cleared area becomes a virtual desert, so more forest is sacrificed to grow more cheap beef.[34] Christopher Uhl, of Pennsylvania State University, estimates that a quarter-pound hamburger derived from steers raised in Central America represents the loss of 55 square feet (5 square meters) of immeasurably valuable tropical rain forest containing one giant tree, about fifty smaller trees, twenty to thirty different tree species, about a hundred species of insects, and many other bird, mammalian, and reptilian species, as well as a huge diversity of fungi, lichens, mosses, and bacteria—in short, millions of living organisms and thousands of species are sacrificed for one quarter-pound hamburger.[35] Similar figures hold true for cattle raised on soil derived from the Amazon jungle.

Illegal Drugs

Cocaine accounts for the destruction of 1.7 million acres of Amazon forest in Peru, or about a tenth of the total deforestation in that country during the twentieth century. For the past twenty years, nearly 6 million acres (about 2.4 million hectares) of tropical rain forest has been cleared in the Andean region of South America. In Colombia, 3 million acres (about 1.2 million hectares) of forest has been destroyed. In Peru, 10 percent of total rain forest destruction in the past century is due to illegal drugs. The United States consumes about 260 metric tons of cocaine annually.[36] The demand for cocaine and crack in the United States is vast, and as long as it continues, there will be suppliers. Furthermore, the traditionally very poor peasants of Peru make a good living growing cocaine.[37] In Bolivia, 38,000 tons of toxic waste, containing forty-one different chemicals, was being dumped as far back as 1991 into tributaries of the Amazon River each year in the process of cocaine production. This practice could turn the Amazon basin into an ecological disaster area. The chemicals included over 309 tons of sulfuric acid,

7,000 tons of calcium sulfate, and 3 million gallons (about 11.4 million liters) of paraffin, which is almost nonbiodegradable and which asphyxiates fish and plankton.[38] Millions of tons of toxic chemical substances and waste by-products continue to be dumped into the Amazon River by cocaine and heroin growers each year.[39]

Pharmaceuticals

We have only begun to explore the natural riches of the rain forest. Thirty percent of all prescription drugs in the United States contain active ingredients derived from rain forest plants. Others are synthesized to resemble these natural chemicals. Seventy percent of the chemotherapeutic agents used in treating cancer are extracted from rain forest plants, as are some antihypertensives, anesthetics, muscle relaxants, and contraceptives.[40] Fourteen hundred rain forest plants contain anticancer chemicals.[41]

Tropical fruits and nuts, oils, rubber and medicines, natural pesticides, and predators of crop-attacking insects offer an enormous and virtually untapped economic resource, potentially two to three times greater than all the timber of the forest. All these products are renewable and sustainable in perpetuity, requiring no fertilizers, pesticides, herbicides, fungicides, machinery, or fossil fuels to produce. Obviously, the distribution and sale of products like these should be seen as the economic wave of the future for the Brazilian government. Saving and nurturing the indigenous forest is far more profitable than logging. The trees could be left standing, and the lungs of North and South America could be saved.[42]

Yet almost half of the Earth's original 4 billion acres (about 1.6 billion hectares) of rain forest has been destroyed.[43]

War

In the past, forests were systematically destroyed in times of war. The US military was responsible for a vivid example of deforestation during the Vietnam War. It sprayed 19 million gallons (about 72 million liters) of Agent Orange over huge areas of rain forest to defoliate the trees and expose the "enemy."[44] Agent

Orange is contaminated with dioxin, one of the most carcinogenic substances known. When I visited Vietnam in 1986, huge mountains still looked like the backs of dinosaurs, blackened and scarred twelve years after the war's end. The only trees that would grow on these poisoned lands were Australian gum (eucalyptus) trees. Indigenous wildlife had been decimated along with the forest, and people were developing tumors in abnormally high numbers. The incidence of birth defects has also increased above normal levels, almost certainly induced by dioxin poisoning.

Other Threatened Rain Forests around the World

Tropical rain forests occupy less than 6 percent of the land surface of the globe and are home to over 50 percent of the world's plant and animal species.[45] They are diminishing in countries such as Indonesia, Borneo, Malaysia, the Philippines, New Guinea, India, Ghana, and Australia. In Malaysia, the survival of a small indigenous tribe called the Penan is under threat as international companies log twenty-four hours a day, at night under huge arc lights. Forests are being "mined," and for every 10 acres (4 hectares) destroyed, only one is replanted. Once the forest is gone, the Penan people may physically survive, but their culture, traditions, and religion will be shattered.[46]

During the rule of President Ferdinand Marcos in the Philippines (1965–86), thousands of acres of tropical forest were sold and logged to provide money for Marcos and his friends and relatives. In Papua New Guinea, a five-year study of satellite images demonstrates that almost 10 million acres (4 million hectares) of rain forest has already been destroyed by illegal logging.[47] In Haiti, which once boasted large areas of forest, only 2 percent of it remains, and the Central American forests that covered 60 percent of the land in 1960 were reduced to 33 percent by 1991, as the developed world built lavatory seats, bookshelves, and houses from this beautiful unrenewable timber.[48] Forty-seven percent of South America's land contributes 22 percent of the world's forest area. Between 1990 and 2005, that continent lost 158 million acres (64 million hectares) of forest land, which accounts for 47–51 percent of the total land area. Forest area has, however, increased somewhat in Chile, Costa Rica, Cuba, and Uruguay.[49] But this is not a sufficient antidote to our quickly diminishing forest. The time to stop this waste is now. The planet cannot afford this reckless destruction.

Acid Rain

Forests are dying not just directly as a result of logging and burning; they are also threatened by indirect dangers, brought on by our way of life.

For example, when flying over the great expanses of Canada, I look down on beautiful lakes that dot the landscape and realize that though many are crystal clear, they are biologically dead. The water's pH (a measure of its acidity) is so low that all living organisms—fish, crustaceans, mollusks, and algae—have disappeared. In addition, the average pH of the world's oceans is decreasing by 0.1 unit annually.[50]

Rain becomes acidic when sulfur dioxide and nitrous oxide gases, which are products of the burning of fossil fuels, combine with atmospheric water vapor to form sulfuric and nitric acids. These acids return to the Earth either as snow, rain, or fog or as particulate matter injected into the air by power plant chimneys. Canadian lakes have been most severely polluted by acid rain originating in US factories and power plants. Very high chimneys have been built in the midwestern United States to ensure that the pollution does not fall locally but, instead, travels thousands of miles in the wind currents before being deposited on distant lands.

Acid rain has devastated forests around the globe. In New England, the production of maple syrup has been halved in recent years because acid rain is killing the trees it comes from. Fifty percent of Germany's Black Forest has been killed, and similar figures pertain to the East Coast forests of North America, to other European forests, and to the forests in China and South America. Sixty-seven percent of the trees in Britain were sick from these effects in the late 1980s. According to the European Commission's report in 2000, though, 41 percent of Europe's trees are classified in the "warning stage" and 20 percent are damaged.[51] Eighty percent of the lakes in Norway are dead because of acid rain. At the southern tip of Norway, acid rain has killed salmon in all of the larger salmon-bearing rivers.[52]

Since 1980, however, scientists with the United Nations Economic Commission for Europe (UNECE) Convention on Long-range Transboundary Air Pollution have reduced the levels of acid rain in Norway, as well as Europe as a whole, by over 50 percent. Hence, the fishing quality and quantity have greatly improved. Europe (excluding Russia) has about 477 million acres (193 million hectares) of forests, which is a 7 percent increase from 1990.[53] The amount of forest resulting from conservation efforts increased 100 percent between the 1990s and 2005.

Even the Arctic Circle is now covered with a brown haze of smog, and acid rain and snow—originating in European, Soviet, and Chinese factories—that are polluting the lakes, rivers, and trees of Alaska. No country is immune from pollution, because acid rain does not respect national boundaries or artificial lines drawn on maps.[54]

As we have seen, although naturally formed ozone is necessary in the stratosphere to protect the biosphere from ultraviolet light, human-made tropospheric ozone is deleterious to plants. This lower ozone layer is formed when sunlight reacts with car and factory exhausts (smog). In addition to aggravating asthma and chronic respiratory disease in human beings, ground-level ozone damages crops and trees by inhibiting photosynthesis. Unfortunately, the harmful effects of ozone potentiate the plant damage caused by acid rain. So forests and trees suffer even more severely than would be expected, because the detrimental effect of one chemical enhances that of the other. The average levels of ozone in the cities of Europe and North America are three times above the minimum level that causes crop damage. Scientists believe that low-level tropospheric ozone has contributed to the death of 87 percent of the ponderosa and Jeffrey pines in southern California.[55] Between 2005 and 2006, scientists monitored the mortality and air pollution in the San Bernardino Mountains of southern

California. Several of the sites indicated that all pines had died during the drought of 2001–4, which accounts for 31 percent of the area. In contrast, few pines had died in areas with low air pollution, even though little precipitation had been present.[56]

In 1990, Congress attempted to address the acid rain problem by issuing major reductions in sulfur dioxide emissions in the Clean Air Act. Emissions have been reduced by 25 percent since then, but eastern lakes in the United States and southeastern Canada have shown little or no improvement. Forests in these areas, especially central Ontario and the Atlantic Provinces, have suffered greatly, because they also absorb the acid rain that feeds the lakes. Legislation is needed to reduce emissions of sulfur dioxide and nitrous oxides more drastically than the Clean Air Act did.[57]

Living in affluence does not necessarily make us happier than we would be if we lived close to the land, grew our own food, rode bicycles, and read by candlelight. After all, Dickens, Shakespeare, Beethoven, and Brahms wrote by candlelight. I am not suggesting we all need to do this, but some people may want to. Conservation and alternative energy are certainly in order. Television should be almost discarded because the endless advertising tends to encourage this destructive lifestyle; in any case, we learn very little from most TV programs. Better to indulge in family dynamics, to play cards and chess, to sit around the piano singing and playing, than to curl up and numb our brains, our feelings, and our powers of critical thinking by watching the boob tube. However, if television were used in a responsible fashion by the corporations who own this medium, it could become a wonderful educational tool. Actually, the public owns the airwaves; it's just that corporations have been allowed to co-opt them for their own ulterior motive: profit making.

Paper Production

The production of paper has had a devastating effect on the forests of the world. But before we discuss paper production itself, let's talk about the role of paper in today's society. When I was a little girl, newspapers were small in bulk and packed with news. Now, newspapers are bulky because they carry thousands of classified ads and full-page ads for automobiles and the like. The news coverage is sparse, and many papers resemble TV journalism, for the photos are colored and the news bites are one or two paragraphs with no background information. The *New York Times* weighs several pounds on Sunday, full of advertising that very few people read. And very few newspapers are printed on recycled paper. After being read, some newspapers are recycled, but many are sent to garbage dumps, there to be burned and converted back to carbon dioxide or to rot gradually over many years. Recycling a single run of the *New York Times* would save seventy-five thousand trees.[58] One ton of recycled paper saves seventeen trees, 380 gallons (about 1,400 liters) of oil, 3 cubic yards (about 290 cubic meters) of landfill space, 4,000 kilowatt-hours of energy, and 7,000 gallons (about 26,000 liters) of water.[59] If all Americans recycled one-tenth of their newspapers, about twenty-five million trees would be saved each year.[60]

Newsstands are really magazine supermarkets filled with journals on diverse subjects ranging from pornography to knitting, auto mechanics, boating, racing, gardening, fashion, and cooking. Most of the glossy pages are covered with advertisements; only one-third to one-half of the paper contains articles. Glossy paper for such publications is produced from short-fiber hardwood, including eucalyptus trees from Australian forests. Until recently, this timber was thought to have no economic value because it is extremely hard and cracks and splits, and because in many cases it is almost impossible to drive a nail into it. But over the last twenty years these forests have become very lucrative, and the Japanese have moved in to log Australia's last remaining wildlife habitats. In December 2000, the Consumers Union of Japan (CUJ), which represents Japanese consumers with environmental concerns, announced that it did not support the supply of wood chips from Australia's old-growth forests to Japanese paper manufacturers. The significance of this announcement is that the CUJ represents many wealthy individuals who purchased almost all of Australia's wood chips in the past. The CUJ also met with members of the Japanese Parliament to curtail deforestation in Australia.[61] Suffice it to say that these actions, although noteworthy and noble, have had no significant impact, and Australia's beautiful forests are still being decimated for Japanese paper.

We must become conscious of the profligate use of paper. Consider the junk mail that fills your mailbox each day; send it back to its source. You will find that the volume of this useless mail will soon decrease, and paper will be saved.

We blow our noses and wipe our tears on trees instead of on cotton. Handkerchiefs, reusable for years, are hygienic and eminently sensible in these days of forest degradation. Instead of paper towels in the kitchen and paper napkins at dinner, reusable cotton towels and table napkins are better choices.

Tampons are made from trees and packaged in plastic applicators. Discarded in the sewage system, they can wash up on beaches and last for five hundred years. Tampons can be efficiently inserted by means of the fingers. A useful alternative is a cosmetic sponge, which can be inserted dry, removed every few hours, thoroughly washed and squeezed dry, and then reinserted.

Babies are swathed in white paper diapers covered with plastic. Twenty-one billion disposable diapers are used annually in the United States alone.[62] When I was twenty-seven, I had three babies under the age of three years, and they wore cotton diapers that I washed and hung on a line in the sun to dry. They smelled wonderful, and I liked nurturing my babies by washing their diapers. From a public health perspective, it is significant that feces are disposed of down the toilet but paper diapers full of human excreta are sent to the dump, where pathogenic bacteria could well contaminate drinking-water supplies.

Trees are victims of overpackaging. Think of how tea is served these days in paper bags. (The tea bags fizz when placed in hot water; have you noticed?) Often, in classy restaurants, the tea bags are packaged in aluminum foil bags and arranged in boxes made of beautiful wood. (Aluminum is manufactured from bauxite ore. Large quantities of electricity are used to smelt bauxite, thus adding to the greenhouse carbon dioxide.) The boxes are probably cut from rain forest timber. It is quite unnecessary to put tea in all this packaging. Loose tea leaves should be placed in a teapot and covered with boiling water; the tea leaves sink to the bottom of the pot. Tea has been made like this for generations.

Tea bags are but one example of grossly excessive packaging. When I was young, I used to ride my bicycle to the grocer's shop and ask for one pound of sugar. The nice grocer ladled the sugar from a big jute bag into a brown paper bag, which he placed on a pair of scales. He weighed the sugar and closed the bag securely, and I rode home, mission accomplished. No cars, no greenhouse gases, and no superfluous packaging.

A visit to the supermarket today offers a lesson in redundancy. Every item is packaged and covered with plastic, paper, cardboard, or foil—often many times over. Packaging companies make large fortunes as we unthinkingly purchase their products. The next time you shop, take your baskets and string bags, and at the checkout counter unpackage the eggs, tomatoes, and sugar down to the paper bag, placing them in your baskets and leaving the packaging strewn over the counter. Encourage your friends, relatives, and neighbors to do the same, and soon the supermarket management will be buying goods in bulk and discarding packaged goods to foster customer satisfaction.

Printers and computers eat paper and use it in a most inefficient way. Businesses and universities stress communication, but instead of talking to one another, participants cover reams of paper with redundant and unimportant memorandums. This wasteful use of technology must change. E-mail has been said to save paper, but in fact huge amounts of paper are still used to print the material received in e-mail messages.

To retain our precious forests, all paper must be recycled. Newspapers and magazines must cut down on advertising, they must be printed on recycled paper, and we must be discriminating about buying unnecessary and superficial magazines and papers.

The production of paper is itself environmentally dangerous. Many toxic chemicals are made and used during the manufacturing of paper, and paper mills smell of rotten eggs. Fishermen in Canada strongly advise Australians not to allow the construction of paper mills, because the effluent destroys the surrounding countryside and kills the fish. I drove through the town of Lewiston, Maine, recently. There we crossed over a bridge that spanned a large river and upstream waterfall. This sight must once have been a tourist attraction, but the water splashing over the rocks was green-brown and smelled foul. It emanated from several paper mills upstream. I was surprised to learn that almost all the forests in Maine are owned by private paper companies and are therefore in peril. A local conservationist told me that this is true in most of the beautiful New England states, which derive a large part of their income from tourists who are drawn to walk through the magnificent golden and crimson forests in October and November.[63] I wonder how many New Englanders understand how precarious this natural beauty is.

Paper production has been shown to be medically dangerous to workers. Although paper pulp can be bleached by oxygen or ozone processes, chorine is still used extensively to bleach paper. An analysis of 1,071 deaths among paper mill and pulp workers in New Hampshire from 1975 to 1985 showed an increased incidence of cancer of the digestive tract and lymphoid system.[64] These workers were occupationally exposed to many dangerous organic chemicals, including dioxin, that are released into rivers and lakes, and that pollute the water and concentrate in fish. In August 1991, interestingly, EPA Administrator William Reilly issued a report indicating that dioxin is not as carcinogenic as was originally thought, and that the EPA had opened a yearlong evaluation of the risks associated with dioxin. However, Ellen K. Silbergeld, professor of pathology at the University of Maryland at Baltimore, strongly disagreed, saying, "Nothing that has been learned about dioxin since 1985 when the EPA first published its risk assessment finding on dioxin in the environment supports a revision of science-based policy or action."[65]

In 1997, the International Agency for Research on Cancer (IARC) announced that dioxin (2,3,7,8-TCDD) is considered a Group 1 carcinogen—that is, a "known human carcinogen." In 2003, another cancer report confirmed that there is no known "safe dose or threshold" below which dioxin will not cause cancer. Other environmental scientists have suggested that, in its report, the government was attempting to protect not people, but the industries that manufacture dioxin.[66] Dioxins are not only carcinogenic, but they can also induce harmful reproductive and developmental problems. Furthermore, they can damage the immune system, interfere with hormonal mechanisms, cause birth defects, reduce sperm counts, induce learning disabilities, cause diabetes, bring about lung problems, and lead to skin disorders, to name just a few examples.[67]

Once dioxins enter the environment, they are easily distributed great distances. Dioxins decompose slowly, whether in water, soil, or plants; and they accumulate in the food chain—for example, in the fatty tissues of animals after ingestion. Consumption of contaminated animals, in turn, exposes humans to low-level concentrations of dioxins. The good news is that dioxin levels in the United States have been declining for the past thirty years. Thanks to the EPA, state governments, and industry, dioxin levels have diminished more than 90 percent from 1987 levels. As a result, human body burdens are in decline.[68]

It is possible to make paper from fibrous plants such as bamboo, hemp, papyrus, and kenaf. These crops can be grown on degraded farmland and pastures, and forests can thus be saved. With appropriate timelines, trees can also be grown as crops, harvested, and replanted. Reliance on timber for building is appropriate in one sense, because the carbon in lumber is not released as carbon dioxide, but this wood must be taken not from virgin forests, but only from timber plantations. We must explore other building materials, such as mud bricks, that are made from soil adjacent to the building site and dried by solar power. They are widely used in Australian and New Mexican adobe houses, with beautiful results.

Reason for Optimism

Reading this litany of tragedy may be depressing and engender a feeling of helplessness. So let me end this chapter by providing some inspiring examples of human courage and heroism.

Tasmania is a small island state located off the southeast corner of Australia. It is heavily timbered and has a small population. The forestry department has for many years conducted logging operations in the forests, and the hydroelectric commission is fond of creating dams and flooding large areas of bushland. Some of the world's tallest flowering plants (trees) grow in this state.

In the early 1980s, the Tasmanian state government decided to dam a large section of the Franklin River, which covers one of the most beautiful ecosystems in Tasmania. The people of Australia were incensed. Blockades and sit-ins were organized, and the area was occupied for months. The weather was often bone-chillingly cold, but people from all corners of Australia came to demonstrate. The actions were efficiently coordinated, and the Forestry Commission workers never knew what would greet them as they moved their bulldozers and trucks into wilderness areas. Two-way radios, tents, wet-weather gear, food, and other equipment were donated by sympathetic people around the country. One of the outstanding heroes during this operation was a medical doctor named Bob Brown, who, dressed in a suit and tie, sat in a tree for several weeks and gave erudite and inspirational interviews to the press. Eventually, the people of Australia prevailed, a new national government was elected because of the protest, and the Franklin was saved.

Note that battles like these never end. Over the past two years, a large Tasmanian logging company called Gunns has been in league with the Tasmanian government, attempting to build a pulp mill in a beautiful place called the Tamar Valley, which is home to many organic farms and vineyards. The mill will be fed by hundreds of thousands of beautiful old-growth trees, which, over time, will release huge amounts of global-warming CO_2, which has been trapped in these trees for hundreds of years! Like the battle to save the Franklin, a pitched battle to save the Tamar is now in full swing.

Individuals can lead their countries and create movements to save the world. If you are sufficiently inspired to be one of these leaders, all you need is an extensive knowledge of the pertinent facts and a clear determination that you will prevail to save your community and, indirectly, your country and the world. Maintain your naïveté and determination, never procrastinate or become cynical or self-doubting, and you will win every time. We need to plant billions of trees to save the atmosphere. Let's cover the Earth with trees and bring back the birds!

5 Toxic Pollution

When I was a child, women wore silk stockings and silk, rayon, or cotton underwear; floors were covered with linoleum or woolen carpets; china plates, metal cutlery, and glass were used in the kitchen; and garbage consisted of food scraps wrapped in old newspapers. The trash can was a small, lidded, metal can that was put out weekly for collection, and once a week the iceman placed a block of ice in the top of our ice chest, which kept the food cold. The milk was delivered at five o'clock each morning by a milkman driving a horse and cart. Fresh bread was delivered at lunchtime, and Dad grew most of our fruits and vegetables. On the weekends, my father walked the streets with a bucket and spade collecting horse manure to fertilize the garden. Our life was healthy and relatively free of artificial chemicals.

When I was eight, someone showed me a "nylon" doorknob. It looked like a magic material, and intuitively I knew it heralded the dawn of a new era—the plastic age. With some trepidation, I watched the introduction of Laminex laboratory benchtops, vinyl floor tiles, plastic lavatory seats, plastic-covered furniture, plastic piping for plumbing, plastic-filled planes and cars, disposable plastic kitchen utensils, and plastic medical equipment. The gravity of the situation became apparent when I was a medical resident in 1972. I asked a hospital administrator why we used disposable plastic scissors, scalpels, forceps, and so on, instead of metal surgical instruments. He told me it was cheaper to buy plastic and throw it away than to employ people to wash steel implements.

Forty years later, the situation is totally out of hand. I have witnessed the takeover of the world by plastic. In the hotel rooms I stay in on my travels, there is a plastic sachet full of nonbiodegradable shampoo, a plastic shower cap, and two plastic cups, all wrapped in plastic.

Americans and many other people have become obsessed with bottled water. An average of forty *billion* plastic water bottles are used each year in the United States.[1] Or, to put it in another context, Americans drink about 8 million gallons (about 30 million liters) of bottled water per year, and they spend $11 billion on water, making bottled water the fastest-growing segment of the beverage industry.[2] Marketers have mastered the art of selling water in plastic bottles by using phrases such as "fresh alpine water" or "pure glacier water." But experiments with students reveal that they cannot tell the difference between bottled and tap water imbibed from unmarked glasses.[3] Bottled-water manufacturers can print these phrases as long as they comply with FDA regulations that the water does not come from the surface of a well, although what difference that makes to the quality of the water is not clear. Some water-bottling companies take water out of springs, and some filter ordinary community municipal sources. Some bottled water has added minerals that make absolutely no difference to the health of the drinker.[4] Soon, I suppose, private companies will work out how to sell air! I cannot really imagine how people can make billions of dollars selling water—*the ultimate scam*! The public is ultimately susceptible to false and seductive advertising.

Over 1.5 million barrels of oil each year is required to produce these billions of plastic bottles—enough oil to fuel a hundred thousand cars. More than 85 percent of the plastic bottles are not recycled, and it takes more than a thousand years for the plastic to biodegrade, if it ever does.[5] In fact, it takes about a quart (1 liter) of oil to produce about 5 quarts (5 liters) of bottled water, including the manufacture of the bottles, transportation, and refrigeration—all energy-intensive activities. Interestingly, but not surprisingly, the Australian beverage supplier Coca-Cola Amatil owns and produces much of this highly lucrative bottled water.[6]

Meals on airplanes demonstrate our "throwaway" society. Food is served on plastic plates covered with clear plastic wrap (a substance that emits a very toxic carcinogenic component called vinyl chloride, particularly when heated in a microwave oven). Tea is served in Styrofoam cups, which were foamed with CFC gas and which may contain untreated styrene, a carcinogen. These containers last at least five hundred years. Milk is presented in a white cardboard carton contaminated with dioxin, and all the food tastes strangely of chemicals. Following the meal, the hostess walks down the aisle collecting all "disposable" utensils in a big plastic bag, which will be deposited in a garbage dump. About 25–30 percent of the volume of municipal garbage is plastic. Over 245 million tons of municipal rubbish (or 4.5 pounds [2 kilograms] per person per day) was dumped into US landfills in 2005. Paper, the most common material thrown away, represented 34.2 percent of landfill material.[7] Plastic is not biodegradable, and it remains in its original form for a millennium.[8] Everything we use is eventually buried in the soil—plastic, cars, refrigerators, houses, tables, diapers, even our own bodies. I wonder how future archaeologists will classify our particular form of civilization when they analyze our garbage. And another point: "landfill" is a convenient and dishonest euphemism for garbage.

Most plastic is very difficult, if not impossible, to recycle. It is synthesized from oil, and scientists have crafted many different chemical molecules to make hundreds of forms of plastic endowed with physical properties like flexibility, strength, and transparency. When these different varieties of plastic are melted, the molecules change their physical characteristics, so it is virtually impossible to recycle plastics in their original form. Although some specific plastics are recyclable, most are not.[9] Fast-food chains demonstrate the profligate use of plastic by modern societies. The volume of packaging far exceeds the volume of food, and almost all this material is either plastic or cardboard derived from trees. Furthermore, fast food is not very nutritious, but it produces a sense of "satiation"—it is so laden with fat that the stomach empties slowly and one feels full and satisfied for longer periods of time than if one had eaten healthy proteins and carbohydrates. Fat is cheap, so it is economically advantageous to sell potatoes (carbohydrate) and hamburgers (protein) soaked in fat. Since 2003, entrée-sized salads have been added to some fast-food menus, as well as chicken sandwiches, fruit and yogurt, and soup. Yet McDonald's generates three hundred tons of rubbish daily at its thousands of restaurants in the United States—most of it paper and cardboard packaging made from trees. McDonald's has "metastasized" into thousands more franchises in other parts of the world,[10] and the average McDonald's restaurant serves two thousand customers a day, generating 238 pounds (108 kilograms) of waste daily. This is simply obscene.[11]

In the manufacturing of plastic, large quantities of different chemicals are made as by-products. They constitute toxic waste that is emitted into the air through factory chimneys; poured into sewage systems; sent to garbage dumps; drained into streams, rivers, and lakes; buried in landfills; or illegally dumped at night. The food chain concentrates many of these toxic organic chemicals, so the plastic we use every day may come back to haunt us as poisonous food, water, or air; or it stays in our garbage dumps for hundreds of years.

There are other interesting ways to dispose of toxic waste; the following story describes one of the more innovative approaches. An attorney, Mark Pollack, twenty years ago attended one of my speeches, in which I addressed the medical effects of nuclear war. He left the lecture saying to himself, "Well, I'm either part of the problem or part of the solution." He decided to create a new legal specialty and opened a practice in which he concentrated on toxic-waste issues. Pollack sues firms for willful and knowing contamination of the environment. For example, he sued Shell Oil for dumping toxic wastes into San Francisco Bay and won $18 million.

Early in 1990, a woman in Napa Valley called Pollack. She was worried that a lake near her house had turned blue. He told her not to be ridiculous—that lakes were not blue—but because she was insistent, he decided to investigate. The lake was indeed blue. He noticed that adjacent to the lake stood a large building. The man who opened the door when he knocked had blue material running from his nostrils. He said he manufactured a blue chemical, some of which he drained into the lake. As Mark stood in the office, he noticed two lots of large barrels stacked in the corners of the room. He asked what the barrels contained, and the man informed him that he had been paid to store them by chemical companies who wished to dispose of their toxic waste. On further questioning, he became quite enthusiastic and said he mixed together the two chemicals contained in those barrels to make a thick paste. He then sold this paste to certain mortuaries, where it was stuffed into the thoracic and abdominal cavities of cadavers after autopsy, during which the organs are removed. Thereupon, the bodies were dressed in their best clothes and buried after appropriate funeral rites. These bodies then became repositories of toxic waste, and the disposal problem of the chemical companies was solved. This anecdote may give a clue to many other illegal means of waste storage.

A young woman in Los Angeles said to me the other day, "It's a good day today; you can see through the air!" The people of Los Angeles remind me of the classic frog experiment in which a frog is put in cold water and the water is gradually heated. The frog does not notice the temperature change and eventually boils to death. If you drop a frog into boiling water, however, it leaps out and saves its life.

A recent study performed in Los Angeles at the University of Southern California showed that in the lungs of one hundred youths who had died violent accidental but nonmedical deaths, 80 percent had "notable abnormalities in lung tissue, and 27 percent had severe pathology." The pathologist, Russell P. Sherwin, who performed the study, said, "The youths were running out of lung . . . Even if I were to assume that most of these people were smokers, I'm seeing much more damage than I would expect to see."[12] In January 2005, Breathe California of Los Angeles County (BREATHE LA)[13] inaugurated the Center for Healthy Lungs. The center teaches children to understand alternative fuels, health effects, and sources of air pollution and to promote clean air in their communities.

About 125 million pounds (57 million kilograms) of toxins are discharged into the Los Angeles air each year, including the potent carcinogens trichloroethene, methylene chloride, and benzene. In a start toward reducing pollution, in 2005 13,000 tons of benzene was discharged into the air in California—40 percent less than in 2001. Butadiene emissions declined as well, to 3,000 tons—60 percent of the 2001 level. The toxin levels are still too high; one in every 714 residents of Los Angeles or Orange County has the potential to develop cancer as a result of air pollutants—twice the national average. According to EPA data, benzene is responsible for a risk of twenty-four cancers per million people exposed, and butadiene is responsible for ten cancers per million.[14] In 2004, according to a Sierra Club report, 47 percent of LA's smog came from the exhaust of cars and trucks. Diesel exhaust accounts for 70 percent of the total cancer risk in the LA basin.[15] Each person generates 65 pounds (about 30 kilograms) of smog per year.[16]

Pesticide fumes in California came under regulation in May 2007—a move that should have been implemented ten years earlier. The chemicals are poisonous gases that sterilize the soil in order to kill insects, weeds, and diseases. But obviously any chemical that kills biological systems also affects human biology. After evaporating from the soil, the gases cause smog. In 2005 alone, about 36 million pounds (16 million kilograms) of fumigants were sprayed on California farms; and these fumigants linger on the food. Farmers, such as strawberry, carrot, tomato, and pepper growers, are between a rock and a hard place, for they stand to lose an estimated $10 million to $40 million a year because of these new environmental laws. And if they cannot afford to keep their farms, they will be forced to sell to "developers," who will cover wonderful farmland with houses, concrete, and asphalt. California is the first state to regulate these chemicals.[17] Obviously, we as a society must turn to organic farming. The overall profits will not be as great, but at least the food will be fit to eat.

Many corporations seem to be hopping onto the green-movement bandwagon, but caution is indicated because some skirt the truth. For instance, British Petroleum paints its gas pumps and oil tankers bright green, so if a tanker accidentally spills millions of gallons of crude oil into an environmentally sensitive bay, the general public should be reassured because the ship was painted green. This comment is facetious but basically true because the true motives of BP are to continue business as usual selling oil as fast as they can, camouflaged by some greenwashing.

I am also skeptical about "environmentally friendly" products that appear on supermarket shelves. The list of chemicals often contains toxins. Even if the product seems to be chemically innocuous, often large amounts of electricity were used in its production. The ecological claims made on the labels tend to be fraudulent—degradable plastic, recycled plastic, "decomposable" diapers, and so on. Plastic is not biodegradable, and it cannot be recycled into the original chemical. The plastic that covers diapers does not decompose.

To date, seven million artificial chemical compounds have been synthesized.[18] Most of these have been noted only once in the chemical literature. More than 2 billion pounds (about 900 million kilograms) of toxic chemicals are used for US crops, lawns, and gardens each year.[19] Many are released deliberately into the environment precisely because they are toxic—to kill weeds, trees, and insects. More than $13 billion worth of pesticides are used globally each year.[20] In 2005, $8 billion worth of nonagricultural pesticide toxins were used globally. The leading companies that produce these toxins are Bayer, Dow, FMC, Syngenta, DuPont, and BASF.[21] In 2004, total global pesticide sales reached a record $15.9 billion.[22] About eighty-two thousand chemicals are now in common use, almost all of them toxic.

We continue to add more than two thousand new chemicals each year to this frightening inventory, and very few are methodically tested for carcinogenicity or mutagenicity.[23] The EPA reviews an average of seventeen hundred new chemicals that the industry is seeking to introduce each year, and it approves about 90 percent of them without restrictions. Out of the eighty-two thousand chemicals in use, only a quarter have been tested for toxicity levels.[24] In 1991, no information was available on the human toxicity of 63,200 of these commonly used chemicals, and a complete toxicology profile had been prepared for only 1,600.[25] All of them were derived from scientific research and development. Many are necessary components of industrial production, and others are unwanted by-products. Only a thousand out of eighty-two thousand chemicals on US shelves have been tested on animals for carcinogenicity or their ability to mutate DNA. The tests for each chemical cost about $2 million.[26]

The number and quantity of chemicals in common use boggle the imagination. The chemical industry produced seven times more goods in 1990 than in 1960, and global production of organic chemicals increased from 1 million tons per year in the 1930s to 250 million tons in 1985. The United States accounts for 26 percent of global chemical production, making it the largest chemical producer in the world. In 1997 the US chemical industry developed an estimated $389 billion worth of products and exported about $71 billion worth of chemicals.[27] In 1961, global use of pesticides was about ½ pound per acre (½ kilogram per hectare), but by 2004 it had reached nearly 2 pounds per acre (about 2 kilograms per hectare).[28] Today the industry provides more than $163 billion to the US GDP, or 2 percent of the total.[29] Annual production is now doubling every seven to eight years.[30]

Because of our own addictive consumption, people in the developed world cause a hundred to a thousand times as much pollution per capita as do people in the developing world. (The US population, a mere 4 percent of the world total, creates half of the world's toxic waste. The average US citizen affects the world ecology twenty to a hundred times more negatively than does the average person in a developing country.)[31] Of the 350 million–500 million metric tons of hazardous waste manufactured on the planet each year, the United States produces 245 million–260 million metric tons—almost one ton per person, or about 4.5 pounds (2 kilograms) of waste per person per day.[32] This pollution is a natural consequence of slick advertising, lack of community education on the part of the scientific community, and an extremely high standard of living, which contributes over time to much sickness and suffering. Add to this indignity 6.5 billion tons of carbon emissions per year from the burning of fossil fuels—four times the 1950 levels[33]—and in 2004 an estimated 2,000 metric tons of high-level radioactive waste in the form of spent fuel was produced in the United States alone.[34]

Dangerous chemicals are found in almost every household article. Carcinogens are used to dry-clean clothes, and traces of chemicals often remain on the garments when we collect them from the shop—you can actually smell them. Paints and thinners are supplemented with toxic chemicals designed to discourage fungal growth or act as quick-drying agents. Cancer-causing chemicals are used to kill termites, cockroaches, and insects in homes, offices, schools, hospitals, and restaurants. Cleaners and deodorizers contain toxic chemicals, as do garden pesticides and fungicides. Toxic fumes emanate from synthetic carpets, furnishings, and curtains. Carcinogenic formaldehyde leaches into the air from certain types of insulation and fabrics used in walls and ceilings.

Several commonly used chemicals need to be described in some detail. The clear plastic bottles that are used for soft drinks, baby bottles, and the like are made of polycarbonate plastic, which leaches a specific carcinogenic and hormonal mimicker called bisphenol A into liquids and foods in the plastic container. This chemical has been associated with the early onset of puberty, impaired immune function, prostate cancer, obesity, other types of cancer, and hyperactivity even in very low doses. The Centers for Disease Control and Prevention found bisphenol A in the urine of 95 percent of the people they tested. Other products containing this chemical range from reusable water bottles and microwavable food containers to dental sealants, paints, glues, protective coatings, and the linings of metal cans containing food and drink.[35]

Another very dangerous chemical is perfluorooctanoic acid or PFOA, which is used in the production of Teflon. Made by DuPont, Teflon is the material used to make nonstick cookware. PFOA causes cancer and other health anomalies in laboratory animals and is likely a human carcinogen as well. When heated to high temperatures, Teflon pans emit toxic fumes that can kill a canary in close proximity. Similar chemicals are used in food packaging—microwavable popcorn packets (which release PFOA into the oily popcorn), as well as packaging for French fries, pizza, sandwiches, drinks, candy, and paper plates. It is also used in rugs, shampoos, denture cleaners, car waxes, stain-resistant textiles, and clothing. In one survey, PFOA was found in the blood of 90 percent of all Americans and 96 percent of American children, and US residents have the world's highest levels of perfluorochemicals in their bodies. Unlike dioxin, suspected also to be a carcinogen, PFOA does not break down chemically in the environment.[36] It is constantly leaching from the surfaces it is applied to, and the indoor air is filled with these compounds.[37]

In summary, we live in a sea of chemicals and carcinogens foisted upon us by the ever more powerful chemical companies.

I was unsuspectingly poisoned recently when I painted my chimney with lacquer paint. I used turpentine to wash the paint off the brush and my skin, and the next day I awoke with a splitting headache. By midafternoon, my skin was covered with a raised, nonitchy, nodular rash. I was perplexed about the diagnosis and thought I was infected with a virus until I realized that I had been poisoned by the paint. The turpentine had obviously facilitated absorption of the toxic chemical or chemicals. There had been no warning on the label, and I was horrified to learn that exotic chemicals are added to ordinary household paint. The symptoms took several days to subside.

Physicians are surprisingly uninformed about the medical complications of most toxic products in household cleaners, sprays, paints, and detergents. This subject is not an integral part of the medical school curriculum, yet most bottles of household chemicals carry the warning "If swallowed or inhaled, contact your physician."

I am a physician who is quite aware of the dangers of toxins, but when I was invited by the Australian Cotton Growers' Association in 1991 to talk at its annual meeting about the medical dangers of chemicals used in its industry, I realized I was totally ignorant. I embarked on a three-week intensive study program that became an exercise in frustration. I collected literature from the Department of Agriculture, from the cotton industry, and from the Health Department. When I read the extensive list of organochlorines, organophosphates, dioxin-containing sprays, and other chemicals used as fungicides, herbicides, pesticides, and fertilizers, I was shocked. I knew that many were dangerous, but I found it very difficult to obtain accurate and complete information on the path of these materials in the food chain and on the pharmacological consequences of the chemicals in the human body. It was immediately obvious from an ecological perspective that the soil and adjacent waterways would be poisoned by the sprays, and that insects and birds would be damaged, but I was dismayed to discover that cottonseed oil is the main ingredient of cooking oil and margarine. It is not stated on the packaging. When I called the Australian State Health Department for data on chemical residues in the oil, I was informed that the appropriate tests had not been done. The literature advises mothers to wash new cotton garments three or four times to remove chemicals adhering to the cotton fibers before they use the clothes for their babies.[38]

I told the cotton growers and their wives that their industry appeared to be medically and ecologically dangerous, but that specific information about the chemicals was not readily available to physicians. I told them that if I, an environmentally conscious doctor, did not know much, my colleagues would know even less. They received this message unwillingly because although they knew it to be fundamentally true, they made a good living from growing cotton.

Now let's talk about what happens to these toxic chemicals after we finish using them. Their disposal poses an enormous problem. Garbage was relatively benign in the old days because the use of chemicals was not widespread, but it is now dangerous stuff. Into the garbage we readily toss household cleaners, insecticides, spray cans, mothballs, paint thinners, bleaches, ballpoint pens, floor cleaners, plastics, detergents, oil, dry-cleaning chemicals, window cleaners, dioxin contained in white cardboard and paper, and batteries containing lead and acid, not to mention cars containing heavy metals, plastics, and oil, as well as refrigerators containing CFC gas, to name a few.[39] The US population each year discards 16 billion diapers, 1.6 billion pens, 2 billion razor blades, and 220 million car tires, along with enough aluminum to rebuild the US commercial airline fleet four times over.[40] An average of 2.5 million plastic bottles are thrown away every hour in the United States. The average US supermarket store uses about 60.5 million paper bags per year. Over 20 million Hershey's Kisses are wrapped a day, using 133 square miles of aluminum foil that could be recycled, but most of it is thrown away. The United States uses over 80 million soda cans annually. The average American family consumes 29 gallons (110 liters) of juice, 104 gallons (394 liters) of milk, and 26 gallons (98 liters) of bottled water a year. As previously mentioned, the recycling of one run of the Sunday *New York Times* would save about seventy-five thousand trees. Unfortunately, recycling is somewhat random; some communities are stringent about its application, and some are not. About one billion trees' worth of paper is thrown out every year in the United States alone.[41]

The obvious solution to our dilemma is to use cloth diapers, nondisposable razors, recyclable glass bottles for milk and other drinks (bottles that are not melted down and remade but that are washed and used over and over again), bicarbonate of soda and vinegar for cleaning, ordinary soap instead of detergent, fountain pens rather than throwaway ballpoint pens, and biologically safe pesticides. We must build mass-transit systems and, if necessary, discard small bicycle tires instead of car tires. In other words, we need not stay on this treadmill of addiction to hazardous consumption.

Municipal garbage dumps leach toxins into underground aquifers and nearby rivers and streams, thereby endangering wildlife and human food chains. So dangerous was US municipal waste in 1990 that more than half the hazardous-waste dumps flagged by the congressionally appointed Superfund were municipal garbage dumps and landfills. The other hazardous sites were filled with toxic chemical waste from industry.

A new form of waste is designated "electronic waste," or "e-waste." This term refers to consumer electronics, such as TVs, computers, monitors, audio/stereo equipment, VCRs, DVD players, mobile phones, copy and fax machines, and so on. E-waste also contains numerous toxins, such as lead, cadmium, and mercury, that have detrimental health effects. In 2003 alone, the United States generated 2.8 million tons of e-waste and recycled only 290,000 tons. The rest sits in municipal waste streams. Much US waste is exported to developing countries because of lax environmental and occupational laws in those countries. This practice is called "environmental racism"[42]—a subject I'll return to.

To quote from *The Global Ecology Handbook*, "8 out of 10 Americans live near a hazardous waste site."[43] According to the EPA, "one in four Americans live within four miles of a Superfund toxic site," which is classified as the most dangerous sort of hazardous site. New Jersey has 114 toxic-waste sites. Since 2004, US taxpayers have paid more than $3.8 billion in order to clean up waste sites.[44]

One of the most famous toxic disasters happened at Love Canal, in New York State. In 1945, a Hooker Chemical Company analyst wrote in an internal report that the "quagmire at Love Canal will be a potential source of law suits," and he called it "a potential future hazard." Hooker dumped 40,000 metric tons of toxic chemicals, including dioxin, lindane, and arsenic trichloride, into Love Canal, which emptied into the Niagara River, adjacent to one of the world's greatest natural wonders—Niagara Falls. The company later filled in the canal and donated the site for the construction of an elementary school.[45]

In 1980, President Jimmy Carter was forced to declare a state of emergency when toxic chemicals began oozing from the filled canal into basements of houses and when doctors warned that people could develop cancer, leukemia, and birth defects from these materials.[46] The company was sued for $250 million, the estimated cleanup price. But "cleanup" is a euphemism. How can the toxic waste that entered the Niagara River over the past thirty years be retrieved? How will the people who died, are dying, or will die of chemically induced cancer be compensated? How much do you pay a woman who gives birth to a deformed baby, or the deformed person for a lifetime of disability?

Love Canal is but one of thousands of tragedies about to become manifest all over the country. By September 2007, some 1,569 hazardous sites had been officially placed on the Environmental Protection Agency's National Priorities List,[47] which means that they are eligible for "cleanup" under the Superfund legislation, using federally allocated funds. Every state has at least one of these sites, but New Jersey has 114, Pennsylvania has 93, New York has 86, Michigan has 65, and California has 94.[48]

Unfortunately, the situation is grimmer than the official EPA figures would have us believe. In fact, the US government is the nation's chief polluter, and the US Department of Defense (DOD) is the world's largest polluter, generating over 750,000 tons of hazardous waste per year.[49] Federal facilities discharge almost 2.5 million tons of toxic and radioactive waste without having to report a drop. The EPA does not have the necessary authority to hold the DOD accountable for its actions. Currently, the DOD is responsible for twenty-eight thousand contaminated sites in the United States and abroad. The cost to clean up active and former chemical sites in the United States alone exceeds $90 million.[50] Today the EPA still fails to report accurate information about air emissions from government refineries and chemical plants. The Environmental Integrity Project (EIP) reported in 2004 that the carcinogens benzene and butadiene in air emissions are four to five times greater than the level the EPA has reported. EIP concluded that "in 2004, EPA adopted new rules that actually weakened air emission reporting requirements." The Texas Commission on Environmental Quality (TCEQ) also reported that at least 16 percent of toxic air emissions "have been 'kept off the books.'"[51]

The EPA reported in 2001 that 6.61 billion pounds (about 3 billion kilograms) of toxic chemicals had been released between 1998 and 2001.[52] U.S. PIRG reported that the following chemicals were released into the environment in 2004: 96 million pounds (about 43.5 million kilograms) of chemicals linked specifically to developmental problems in children; 70 million pounds (about 31.8 million kilograms) of carcinogens from chemical and paper industries; 826 million pounds (about 375 million kilograms) of neurological toxins; 1.5 billion pounds (about 680 million kilograms) of respiratory toxins. In December 2006, the Bush administration ruled to limit the number of reports released to the public about the quality and quantity of toxic chemicals. Companies are no longer required to submit detailed reports of "bioaccumulative toxins in amounts under 500 pounds." All other waste discharged into the air and water—from 500 to 5,000 pounds (227–2,300 kilograms)—must be reported. Consequently, the general public is bound to know less about the materials that have been and are being released into their communities.[53] As a physician, I find the aforementioned situations extremely disturbing. The incidence of cancer is increasing for obvious reasons: we cannot cure many cancers, and we have a government so influenced by industry that it allows carcinogenic and toxic chemicals to continue to be released into the environment. This situation is totally out of hand and requires urgent intervention by the medical profession and the Obama administration.

Huge fuel storage tanks buried at gas stations are also sources of a disturbing amount of underground pollution. Constructed of steel, these tanks inevitably rust over time, eventually leaking their contents into the ground. It takes just two pints of gasoline or oil to contaminate several million gallons of drinking water.[54] According to some estimates, at least one-third of the underground aquifer of the United States has been polluted with carcinogenic benzene or chemicals from these and other sources. An EPA report published in January 2007 stated that of the 2.2 million petroleum tanks under inspection, 1.6 million have been closed. In 2006, 641,881 tanks were in use; 464,728 leaks were confirmed; 435,631 cleanups were initiated; and 350,813 cleanups had been completed. The report also said that 76 percent of the underground storage tanks complied with the release prevention requirements; 72 percent complied with the leak detection requirements; and 62 percent of the facilities were now meeting both requirements.[55] This situation is obviously not good enough. We are driving our cars at the expense of our groundwater; much of our drinking water is inevitably becoming contaminated with carcinogens.

Airborne toxins are also a frightening consequence of our modern industrial society. Several vivid examples come to mind. St. Gabriel, a town on the Mississippi River, is host to twenty-six petrochemical factories, which belch at least 400 million pounds (about 180 million kilograms) of chemicals, including the carcinogens benzene, carbon tetrachloride, chloroform, toluene, and ethylene oxide, into the air each year.[56] The Lower Mississippi River industrial corridor is known to environmentalists as "Cancer Alley" because over 125 chemical companies have polluted the underground aquifers. Almost three-fourths of Louisiana's population uses these aquifers for drinking water.[57] I have visited this area, which, apart from the factories, is inhabited mostly by poor African American families. This is environmental racism at its worst.

Another example is the aftermath of Hurricane Katrina in New Orleans. This environmental catastrophe caused major damage at many chemical facilities, as well as at Superfund toxic-waste sites across the Gulf region. Funding to clean up the toxic pollution was cut short by the US federal government, allowing many chemicals to leak into the ground.[58] By 1947, there were 177 refineries and chemical plants in Louisiana, and their numbers continued to grow: to 211 in 1962, 284 in 1981, 320 in 2002. Along the Lower Mississippi River, the number of oil-refining and chemical-processing plants rose from 126 in 1962 to 196 in 2002.[59] This area is therefore very dangerous, because it is a hurricane corridor and an extremely dangerous place to locate these oil refineries and chemical-processing plants.

In addition, half a million pounds (about 227,000 kilograms) of vinyl chloride gas was released in the St. Gabriel region in 1987.[60] Vinyl chloride is an extremely potent liver, lung, blood, and brain carcinogen. It is used in the production of polyvinyl chloride, or PVC, the clear plastic used for bottles containing cooking oil, for plastic piping in plumbing, and for many other purposes. Plastic wrap is also made from PVC. PVC is not stable, and vinyl chloride gas tends to leach out of the plastic bottles and plastic wrap into the food, particularly in hot climates or warm houses and supermarkets, and from plastic containers of food in microwave ovens. Vinyl chloride is fat-soluble, so it is likely to collect in fatty foods such as cooking oil, meat, and cheese that are covered with plastic wrap. In fact, PVC plastic is ubiquitous in everyday life—found in items ranging from shower curtains, soft vinyl toys, children's car seats, auto accessories, raincoats, shoes, and mattress covers. Furthermore, many other items commonly used in the house or as a garment contain PVC.

Besides off-gassing carcinogenic vinyl chloride, PVC off-gasses phthalates, which, despite being reproductive toxicants and suspected carcinogens, are incorporated in the plastic to make it softer and more flexible. The European Parliament recently banned the use of certain phthalates in toys, and one US EPA study found that shower curtains can induce dangerous levels of toxins in the air that persist for months. Phthalates are responsible for the "new car smell." Just one PVC bottle can contaminate a recycling load of a hundred thousand PET (polyethylene terephthalate) bottles.[61] Exposure of pregnant mothers to phthalates has also been linked to small penis size and undescended testicles in baby boys, because this chemical interferes with testosterone synthesis. Cosmetics also often contain phthalates, which can be absorbed by the mother through her skin, through her diet, and by inhalation.[62]

The East Coast between New Jersey and Boston is called the "toxic corridor." When I drive past the eerie petrochemical plants in New Jersey's "cancer alley," I am reminded of a surreal landscape out of science fiction. The air smells foul, there is no sign of vegetation or life, the waterways look like oil, and little wooden houses are scattered among huge masses of pipes and chimneys belching fire; yet real people obviously live and work amid and in these frightful monsters.

Another example of global airborne pollution is the 2 million tons of lead that is released into the air each year. This heavy metal concentrates in kidneys, causing renal damage; and it destroys neurones, inducing brain damage and lower IQ in children and adults.[63] Thankfully, the quantity of leaded gasoline in use in the United States decreased during the 1970s and '80s, and it was fully banned under the Clean Air Act in the United States on January 1, 1996. The pioneering work of my colleague Dr. Herb Needleman, who led the research into childhood neurological abnormalities associated with lead exposure, was partly responsible for the ban.

However, one of the most dangerous forms of lead is found in electronics and computers. Over 40 percent of US "landfill" (garbage dumps) is filled now with consumer electronics. From 1997 to 2004, over 315 million computers were disposed of in landfills—adding 1.2 billion pounds (about 544 million kilograms) of lead contamination to the land.[64] The quantity of lead contaminating the soil and air each year from firing ranges alone amounts to 205 million pounds (about 93 million kilograms)![65]

Twice in less than two weeks in August 2007, Mattel, the world's largest toy manufacturer, recalled over eighteen million of its toys worldwide—nine million in the United States alone. Toys are only a fraction of the goods that may be illegally tainted with lead. There have also been recalls on items such as toothpaste, drinking glasses, candles, and silverware.[66]

Pollution is not unique to the United States. In 1989, I stayed in a chateau in the Ardèche, overlooking the Rhône River valley and the French Alps. Although the weather was clear and sunny, only rarely did I see the Alps, because the valley was so filled with smog that the pinnacles seldom pierced the pollution. On a recent train trip in northern Italy, I turned to the man beside me and said, "Look at that smog," and he replied, "That's not smog; that's fog." Fog is a natural occurrence made from pure water vapor; smog is composed of airborne chemicals that act as a nucleus to attract water vapor. The air is so thickly polluted in Europe that it would be impossible for Renaissance artists to paint their pictures these days. The glorious landscapes and mountains have disappeared. Beautiful, tidy Germany is the same: cows graze in green paddocks amid a gray haze. Europe is disappearing from sight, drowning in its own pollution, the most visible sign of its wealth and good economic performance.

If western Europe is bad, though, eastern Europe is impossible. Almost forty years ago, public health specialists recognized that their population was endangered by industrial pollution, but Communist Party officials quashed concern and public criticism. In some areas in Poland, the air was so polluted that people had to live underground in salt mines in order to breathe. In the industrial region of Katowice and Kraków, human placentas showed high concentrations of lead, cadmium, mercury, and other heavy metals, and these placentas were taken only from healthy babies.[67]

In most Eastern Bloc countries and parts of Russia, the air quality has deteriorated sharply over the last four decades. Very few emission controls were ever applied to factory chimneys, and those that were, tended to be antiquated.[68] The World Health Organization (WHO) confirmed in 2003 that urban air pollution in eastern Europe was still a major concern. WHO estimates that 1.8 million deaths could be avoided every year if better environmental policies were implemented.[69]

Twenty years ago, authorities deliberately did not perform surveys on the affected populations to determine whether pollution was having detrimental effects. In Silesia, Poland, however, it was illegal to grow vegetables because of contaminated soil. The Czechoslovak Academy of Sciences declared that 50 percent of the country's drinking water was unacceptably polluted. In northern Bohemia, home to power plants and chemical factories, infant mortality was 12 percent above the norm. When Nikolai Ivanov, a physician, complained that a nearby copper plant was causing arsenic and lead poisoning in his patients in 1989, he was publicly criticized by a local Communist Party leader. In the industrial cities of Leipzig, Halle, and Dresden, in what was formerly East Germany, death rates from cancer and heart and lung disease were 15–25 percent higher than in Berlin.[70] Since that time, however, the health of the population has improved in eastern Europe. For instance, Prague now has one of the lowest incidences of infant mortality in the world because prenatal care is so well organized.[71] Life expectancy in Germany has increased on average by 2.2 years per decade. Cancer mortality rates have also been reduced overall since 1990.[72]

Solutions

Solutions do not abound. Dumping in soil or water is not the answer. Incineration is becoming a popular approach to the problem of disposal. If the organic chemicals are composed only of carbon and hydrogen, they are converted into carbon dioxide and water when incinerated at high temperatures. However, most toxic compounds emanating from plastics and the like contain chlorine, and some contain heavy metals. When subjected to high temperatures, the chlorine compounds are converted into dioxins, which escape from the chimney together with the heavy metals. Incineration is thus a flawed solution.

Mark Pollack, the lawyer who discovered the contaminated blue lake in California, continues to sue more corporations, and his actions should inspire other attorneys. Instead of suing doctors and each other, they can sue the people who really count—the toxic polluters of our planet.

Portland, Oregon, passed a law in 1990 to ban Styrofoam. The city employed an official, nicknamed the "Styro-cop," to enforce the law. He inspected fast-food outlets, restaurants, nightclubs, and supermarkets, levying fines of $250 for the first offense and $500 for a repeat violation. He was passionate about his job and wisely stated, "To use plastic to drink eight ounces of coffee for two minutes and throw it away where it will take up space forever is absurd."[73]

Basically, though, the only solution to the problem of poisonous chemical waste is to stop making the stuff. Recycling is not the solution, because it gives chemical corporations the green light to continue making more plastics, more aluminum, and so forth. Glass bottles must be returned in exchange for a deposit, washed, and reused. Glass must not be melted down and fashioned into new bottles, because this process produces more global-warming carbon dioxide gas. We must stop buying any food or drink that comes in disposable containers, whether made of glass, plastic, or aluminum. All institutions must use china, glass, and stainless steel kitchen and eating utensils. This ecologically sound policy will also provide employment for a greater number of people. Paper, on the other hand, can and must be recycled. We will have to learn to manage without plastic, quick-drying paints, artificial fibers, cars, CFC-run refrigerators, and the vast inventory of household chemicals and revert to the materials used by our grandmothers. The chores of daily life will be somewhat more time-consuming, but maybe we will find satisfaction and even great joy in a bit of hard work and a sense that we are nurturing the planet.

Food

We must be very conscious of the chemical composition of the food we eat. Our grandparents ate fruit that was not perfectly formed and perhaps contained a few worms, but it tasted delicious. Then farmers began using sprays to kill pests and increase their crop yields. Before long, chemical companies got into the act and agricultural products became a large portion of their business. Initially, the yields of chemically treated crops increased quite dramatically, but over time the predatory insects and parasites mutated and developed immunity to the sprays. So the chemical companies developed new and more exotic poisons to combat this human-induced immunity. Despite these carefully applied chemicals, however, crop losses caused by pests have actually increased since the 1940s, from 32 percent to 37 percent by 1991.[74] Meanwhile, the soil has been contaminated. As a result, worms and naturally occurring microorganisms that form the base of the pyramidal food chain have been killed, and bees and birds that fertilize the crops have been poisoned. In a random sampling of fruits and vegetables in 1995, the US FDA discovered that 60 percent of the fruits and 37 percent of the vegetables had pesticide residue on them. About 2 percent of those fruits and vegetables contained pesticide amounts that exceeded maximum limits established by the EPA. It's a terrible thing to realize that the food we put into our mouths could be poisoning us.

Amazingly, less than 0.1 percent of all pesticides ever reach the target insects. The other 99.9 percent are dispersed in groundwater, lakes, weirs, soil, and the air. Groundwater has been contaminated with more than fifty varieties of pesticides in some thirty US states.[75] In a study on pesticides and water conducted in 1998, the US Geological Survey found that 95 percent of streams and 50 percent of wells close to agricultural and urban areas were contaminated with pesticides. The use of pesticides increased from 300 million pounds (136 million kilograms) annually in 1966 to more than 1 billion pounds (454 million kilograms) in 1981. The United States now uses 1.2 billion pounds (544 million kilograms) of pesticides a year, and 5 billion pounds (2.3 billion kilograms) a year are used globally.[76] The World Health Organization estimated in 1995 that roughly three million cases of pesticide poisoning occur each year (two-thirds of them happening in developing countries), including 220,000 pesticide-related deaths (mostly of people who use or apply pesticides).[77] Pesticides are also used by people as the leading method of suicide, according to WHO. The National Academy of Sciences reports that pesticides in food may cause one million cases of cancer in this generation of Americans.[78] It was predicted in 1998 that, on a yearly basis, one million Americans would learn they have cancer; 500,000 would die of cancer, and 10,400 Americans would die of cancer related to pesticides.[79] All this ecological damage to grow unblemished, artificially beautiful food! I believe it is better to eat a sweet, chemically uncontaminated apple containing a worm than a visually perfect, toxic apple that tastes like cardboard.

Farming must be made safe. Organic farming relying on mulch, compost, manure, and naturally occurring pesticides has been proven to be more productive than chemical farming, but that's true only if the water and soil used to raise the crops are absolutely not contaminated with chemicals. In Australia, organic produce attracts prices 20–100 percent higher than those of chemically grown produce. The United States has similar high prices. This state of affairs will almost certainly change as pesticide production decreases with global oil shortage and people become more and more conscious of the food they are eating and feeding to their families.[80]

The environment, then, is poisoned by industry, by waste dumps, and by agriculture. But we still must examine the transportation of toxic waste, including accidents involving chemicals, sewage, and radioactive waste. This chapter contains a litany of horrifying facts and statistics, but before we can cure a patient we must make an accurate diagnosis. Before we can clean up our lives and cure the planet, we must recognize the extraordinary damage that our lifestyle inflicts on the planet.

Accidents

Accidents that disperse chemicals into the environment are now weekly items in world news. Trucks crash and burn, trains roll over and explode, oil and toxic chemicals leak from sinking ships, and fires and explosions engulf huge chemical factories. The major accidents that have occurred since 1959 are listed in the following table.

Three disasters best illustrate the problem of long-term and far-reaching effects from environmental catastrophes like these. In 1986, the Chernobyl meltdown contaminated most countries in Europe (east and west). Radioactive elements that remain poisonous for hundreds to thousands of years rained down on some of the best agricultural land on that continent. Therefore, we must remember that food grown in these areas may well be radioactive for extended periods and that people eating this food will be at high risk for developing cancer, leukemia, or genetic diseases.

In the same year, a dreadful spill of toxic chemicals into the Rhine River followed a fire in a chemical factory located in Basel, Switzerland. Thirty tons of pesticides polluted the river so much that 500,000 fish and 150,000 eels were killed.[81]

In 1984, Union Carbide accidentally released huge amounts of methyl isocyanate gas from its plant in Bhopal, India, killing thirty-three hundred people and injuring twenty thousand others.[82] Thousands of Indians are still suffering today. Over twenty thousand have died of causes attributed to that incident since the disaster twenty years ago. Amnesty International believes that between 1985 and 2003, fifteen thousand people died (seven thousand to ten thousand died in the immediate aftermath) from being exposed to methyl isocyanate, which had leaked into the air from the pesticide plant. Today, roughly a hundred thousand people are suffering from chronic and debilitating illnesses, and many have not received adequate compensation or medical treatment.[83]

The environmental impact of Hurricane Katrina is still being assessed, but environmentalists know that sediment at the bottom of rivers and other water bodies were collecting chemical toxins and other industrial hazards prior to the hurricane. After

Diary of Disaster

1959	Minamata/Niigata, Japan	Mercury discharged into waterways	400 dead; 2,000 injured
1973	Fort Wayne (IN), US	Vinyl chloride released in rail accident	4,500 evacuated
1974	Flixborough, UK	Cyclohexane released in explosion	23 dead; 104 injured; 3,000 evacuated
1976	Seveso, Italy	Dioxin leaked	193 injured; 730 evacuated
1978	Los Alfaques, Spain	Propylene spilled in transport accident	216 dead; 200 injured
	Xilatopec, Mexico	Gas exploded in road accident	100 dead; 150 injured
	Manfredonia, Italy	Ammonia released from plant	10,000 evacuated
1979	Three Mile Island (PA), US	Nuclear reactor accident	200,000 evacuated
	Novosibirsk, USSR	Accident at chemical plant	300 dead
	Mississauga, Canada	Chlorine and butane released in rail accident	200,000 evacuated
1980	Somerville (MA), US	Phosphorus trichloride released in rail accident	300 injured; many evacuated
	Barking (PA), US	Sodium cyanide released in chemical-plant fire	12 injured; 3,500 evacuated
1982	Tacoa, Venezuela	Fuel-oil tank explosion	145 dead; 1,000 evacuated
	Taft (LA), US	Acrolein released in chemical-tank explosion	17,000 evacuated
1984	São Paulo, Brazil	Gas released in pipeline explosion	508 dead
	San Juan Ixhautepec, Mexico	Gas released, causing tank explosion in liquid petroleum gas (LPG) terminal	452 dead; 4,248 injured; 300,000 evacuated
	Bhopal, India	Leakage at pesticide plant	>2,500 dead; thousands injured; 200,000 evacuated

1986	Chernobyl, USSR	Nuclear reactor accident	>25 dead; at least 300 injured; 90,000 evacuated; fallout spread over much of Europe
	Basel, Switzerland	Fire at pesticide plant	Rhine seriously polluted
1987	Kotka, Finland	Monochlorobenzene spilled in harbor	Seafloor polluted
1989	Alaska, US	Oil spill (*Exxon Valdez* tanker)	11 million gallons of crude oil released into Prince William Sound
	Qualidia, Morocco	Oil tanker (Iranian) explosion	70,000 tons of crude oil spilled, endangering coastline
1990	California, US	Oil spill (US tanker)	300,000 gallons of crude oil released, polluting 14 miles (22 kilometers) of Bosa Chica, one of California's largest nature reserves
1992	Northwest coast of Spain	Oil spill (Greek tanker)	Almost 80,000 tons of crude oil released
1993	Kadar, Thailand	Fire at toy factory	188 women killed; 400 injured
1998	Jesse, Nigeria	Oil pipeline explosion	>500 killed; several hundred severely burned
	Toulouse, France	Explosion at Azote de France (AZF) fertilizer factory	31 killed; 650 hospitalized
2001	Toulouse, France	Second explosion at AZF fertilizer factory	29 killed; >2,500 injured
2004	Glasgow, Scotland	Explosion at plastics plant	9 killed; >40 injured

SOURCE: United Nations Environment Programme (UNEP), "Hazardous Chemicals," UNEP Environment Brief No. 4; George Draffan, "Chronology of Industrial Disasters," Endgame Research, www.endgame.org/industrial-disasters.html.

the storm, the sediment was deposited throughout New Orleans. Tests confirm that high levels of arsenic, lead, dioxin, chromium, and other dangerous chemicals are present. The EPA warned New Orleans residents to avoid contact with the sediment, but other environmentalists believe that some of the chemicals have evaporated into the air, making them impossible to avoid.[84]

As environmental awareness becomes paramount in educated societies, chemical companies that wish to continue their business transfer their most polluting industries and toxic wastes to developing countries. Even Australia has been the recipient of some of Japan's worst polluters. Developing countries are under enormous pressure to accept the unwanted toxic waste of the developed world. In some countries, accepting this toxic waste provides a ready source of income. In 1980, Sierra Leone was offered $25 million by a US firm to store toxic waste. And in the same year, West Germany solicited some North African countries to accept its hazardous waste, and a US corporation attempted to construct a hazardous-waste incinerator on a Caribbean island. Falsely labeled containers of poisonous chemicals are routinely shipped into developing countries, and some of these countries are being encouraged to construct their own waste disposal facilities, which then become the toxic-waste sewage systems of the First World. Lonely ships laden with toxic waste circle the globe for months searching for countries that will accept their cargo, eventually only to dump it surreptitiously at sea. These examples are far from unique.[85]

In 1986 the city of Philadelphia refused to pay Haiti $200,000 (0.008 percent of its annual budget) for the 8 million pounds (363 million kilograms) of toxic incinerator ash that it had dumped onto Haiti's beaches. Ed Rendell, then mayor of Philadelphia, said that the city was too poor to take responsibility for its waste.

The Basel Convention prohibits industrialized countries from dumping their toxic wastes in developing countries. The convention has 170 parties and aims to protect human health and the environment against the adverse effects resulting from the generation, management, transboundary movement, and disposal of hazardous and other wastes. The Basel Convention came into force in 1992.[86] To date, every industrialized country has ratified this convention except the United States. Half of US waste exports are shipped to Latin America.[87]

Accidents with Oil

Oil is pumped out of the ground, and millions of gallons are burned every day in our vehicles and industrial engines. This burning, as we have seen, produces global warming. As we transport the oil to and from refining plants, we spill it from ships and trucks and dump toxic by-products into pits and waste facilities that leach into the underground water. Out of oil, we make plastics and poisons that add to the degradation of the planet. It might seem better to leave oil in the ground, but if we really have to use it, it should be strictly rationed and the price of gasoline should reflect the damage it is wreaking upon our planet and our children's future.

The *Exxon Valdez* spill is a perfect recent example of the oil plague. On March 24, 1989, the *Exxon Valdez* oil tanker foundered on a reef in Prince William Sound, Alaska, spilling 11 million gallons (nearly 42 million liters) of sticky crude oil into a pristine wilderness waterway full of otters, salmon, herring, seabirds, and myriad other wildlife. It was many days before the Exxon Corporation organized itself to begin dealing with the mess, and the local fishermen watched with horror as otters scratched their eyes out and drowned in the oil, seabirds gagged on the toxic sludge, and birds staggered around the beach, their feathers caked in oil. People tried to wash the birds in water and detergent, but most died anyway.[88]

Fifty-eight scientific studies released by the National Oceanic and Atmospheric Administration in April 1991 revealed that the long-term ecological damage of the *Exxon Valdez* accident was far worse than had originally been estimated. Hundreds of thousands of birds and thousands of otters were killed, and baby fish and fingerlings exhibited abnormally high rates of birth defects (demonstrating the mutagenic and teratogenic fetus-damaging effects of the toxic oil products—effects that will probably continue for many generations). Economists who participated in the study estimated the cost of damage to Prince William Sound to be $3 billion to $5 billion, yet Exxon was at first fined only $1 billion. A federal district judge, Russell Holland, later ruled that the fine was inadequate.[89] Exxon paid $2.1 billion in damages up front to clean up shorelines, recover oil, and pay bills, as well as over $3 billion in government penalties. However, it has yet to pay the $5 billion in punitive damages awarded in a 1994 federal court ruling on behalf of thousands of fishermen and many other Alaskans whose livelihoods, water, and land were damaged by the spill.[90] Exxon argues that it should pay no more than $25 million in punitive damages. An Alaskan judge reduced the payment to $4.5 billion plus interest in 2002, but Exxon appealed again and it was later cut to $2.5 billion.[91] On June 25, 2008, however, the US Supreme Court slashed the now $2.5 billion in punitive damages that Exxon had been ordered to pay to a mere $500 million.[92]

Coincidentally, at the same time that the first fine was overruled, Exxon announced unprecedented profits for the first quarter of 1991. They were the highest earned since John D. Rockefeller had founded the company in 1882. Net income rose 75 percent, to $2.24 billion, largely because of sales of jet fuel, kerosene, and other refinery products to the military during the Persian Gulf War.[93] In 2006, Exxon reported that it had earned $39.5 billion, the largest profit reported by any US company in history, and in 2008 it made $11.68 billion in the second quarter, setting a new profit record.[94]

The *Exxon Valdez* incident is not the only ecological violation by Exxon and Texaco, its subsidiary. In 1989, the company was fined $750,000 for failing to perform, and even fabricating, tests of oil-rig-blowout protection equipment. In the same year, it violated the Migratory Bird Treaty Act, which protects certain species of birds. In that year, more birds died in oil company waste pits than had been killed by the *Exxon Valdez* spill. In 1988, a Texaco subsidiary was fined $8.95 million for failing to adequately clean up areas around five thousand corroded, leaking drums of toxic waste.[95] In 1990, the City of New York sued Exxon for submitting inaccurate pipeline safety reports. Exxon admitted that its main detection system had not worked correctly for twelve years, and therefore the company agreed to pay an out-of-court settlement of about $15 million for environmental improvements. In 1996, Exxon paid the Texas Natural Resource Conservation Commission $500,000 because it had dumped 2 billion gallons (about 7.6 billion liters) of chemical wastewater from its refinery in Baytown, Texas. In 1998, Exxon paid $252 million out of the $760 million in punitive damages awarded to Lockheed Martin Corporation workers who had been exposed to chemicals in Burbank, California.

Oil (and its products) is one of the major contaminants of the planet, yet the global community refuses to acknowledge this fact. So addicted are the industrial nations to a steady supply of oil that when Saddam Hussein threatened to control approximately 20 percent of the world's oil supply by invading Kuwait in 1991, the United States and its allies decided to force him out of Kuwait by military action. It is true that the United States had the support of the United Nations and some token military assistance from its allies, but the whole operation was masterminded and organized by George H.W. Bush and the Joint Chiefs of Staff. Bush stated at the start of the Iraqi invasion of Kuwait that the United States was in the Persian Gulf to protect the American way of life. Interestingly, this war came at a time when the Berlin Wall had collapsed, the cold war was over, and the stocks in military-industrial corporations were seriously declining.

Meanwhile, a report from the Organisation for Economic Cooperation and Development (OECD), which represents the world's richest nations, said that the industrialized nations, which make up 20 percent of the globe's population, inhabit 24 percent of its area. Only 18 percent of the world's population lives in the forty developed nations; the other 82 percent lives in the world's 160 developing countries.[96] In 1991, the developed countries produced 72 percent of the world's gross domestic product, owned 78 percent of the cars, consumed 50 percent of the energy (about four times the resources of developing nations), and imported 73 percent of the forest products.[97] Industrialized countries now consume about 175 pounds (80 kilograms) of meat per year, versus about 65 pounds

(30 kilograms) in developing countries.[98] Seafood consumption has quadrupled in the last twenty-five years. The average American consumes five times as much energy as the average global citizen, ten times as much as the average Chinese, and almost twenty times as much as the average Indian.[99] The US contribution to global greenhouse gas emissions is over 20 percent.[100]

People in developed nations, says the OECD report, are intent on worsening pollution of the sea, the air, and the land by ever-increasing "development" (a euphemism for ecological destruction) that encourages the use of the private car and leads to increased congestion and wider dispersion of leisure activities. The amount of waste that each person creates by this lifestyle continues to increase; it now stands at 10 tons per capita per year. The average resident in an industrial country produces on average four to six times as much waste as the average person living in a developing country. Over the past twenty years, this population has become richer, and it lives longer. In the last twenty-five years, per capita consumption levels in this population have risen about 2.3 percent a year, but developing countries are quickly catching up. In East Asia, for instance, the consumption of individuals has risen on average 6.1 percent a year.[101] This wealthy population, of course, includes us. Our selfish and greedy attitudes are now being emulated by China and India, and they seemingly necessitated both of the wars in the Persian Gulf.

I will not discuss the political ramifications of the first Gulf War catastrophe, except to note that hundreds of thousands of people were killed by the most sophisticated weaponry in the history of the human race. The number of Iraqi soldiers killed has been estimated at 150,000; civilian deaths, at anywhere from 5,000 to 100,000. The actual numbers, including the number of children, may never be known.[102] The United States used more firepower in five weeks than was used either in World War II or in the Vietnam War. The air force flew more than ninety thousand sorties—three thousand per day, or one per minute. At least 70 percent of the bombs missed their targets.[103] A UN mission has since reported that Iraq was bombed back to the "preindustrial age" in 1991, and that the resultant situation was "near apocalyptic," with damaged sewage systems and water supplies, few hospital facilities or drugs, and little food.[104] In some ways, to use Pentagon jargon, it was a "near-nuclear war." In truth, it was not actually a war, because there was no retaliation; it was a massacre.[105]

This war produced possibly the most devastating ecological consequences of any other war in history. Immediately after the war, 732 oil wells were burning in Kuwait, set alight or damaged by departing Iraqi soldiers and US weapons that had missed their targets. Red Adair and other oil well fire specialists were brought in to quell the fires, and estimates of the time for completion of the task varied from two to ten years. By mid-September 1991, some 429 of the smaller wells had been recapped.[106] As the fires raged, much of the land in Kuwait and adjacent Iraq, Iran, and Saudi Arabia was covered with a greasy oil slick. Blackened snow was found in the Himalayas, and black soot at Mauna Loa in Hawaii. It is possible that the fires increased global carbon dioxide emissions by 10 percent. Sulfur and nitrogen oxides cause far-ranging acid rain. Indeed, acid rain is "very severe" in Iran, Pakistan, Afghanistan, and the southern part of Russia—which are often downwind from Kuwait. According to the World Meteorological Organization (WMO), the fires daily discharged over 40,000 tons of sulfur dioxide into the atmosphere—an amount equal to the combined emissions of Germany, France, and Britain.[107] In addition, some scientists say that the carbon dioxide emissions produced in the four months after the war ended were equivalent to 1 percent of the normal global production.

Burning oil also produces cyanide, dioxins, furans, PCBs, and other extremely poisonous, cancer-causing toxins and gases, which obviously would affect populations living in the vicinity of the fires.[108] These poisons tend to cling to very fine particles, less than 10 microns (a minuscule fraction of an inch) in diameter, which lodge in the tiny terminal airways in the lung after inspiration. The particles can stay in the lung for years, thus exposing surrounding cells to the cancer-causing chemicals. Thus, oil fires pose potentially a very serious medical problem. In fact, the WMO estimated that one year of burning oil wells could double the quantity of these fine particulates in the world's atmosphere.

The fragile desert ecosystem of the Middle East was damaged by the war, as was the extensive irrigation system of dams, dikes, and channels in the fertile Tigris and Euphrates river systems. Baghdad was one of the most ancient and beautiful cities on Earth, and Iraq itself has attracted archaeologists because it is full of valuable ruins from ancient civilizations. Baghdad was bombed daily, some say even carpet bombed. Antiquities were destroyed in the name of power, oil, and control.[109] It is 2008 as I rewrite this book, and tragically the situation has not improved in Iraq since 1991—in fact, it has become much worse.

A quantity of oil twenty-four to thirty-two times that of the *Exxon Valdez* spill was emptied into the Persian Gulf waters by Hussein and probably by US bombs damaging oil facilities. This oil affected important desalination plants that line the Gulf and supply eighteen million people with fresh drinking water. Three of these plants had to be closed down as the oil slick spread.[110]

Apart from the threat to the desalination plants, the most important side effect from the oil spill was the projected death of hundreds of thousands of seabirds and fish, of which the Persian Gulf has a rich variety. The feathers of seabirds have been gummed up by the oil, and toxic hydrocarbon chemicals dissolving into the water from the slick could threaten turtles, dolphins, whales, sea cows, and other fish and birds.[111]

According to the World Wide Fund, the Persian Gulf is a major migratory flyway for millions of waterbirds, many of which are now being threatened. Information available four months after the 1991 war suggested that thirty thousand to sixty thousand birds had died. The spill also threatened some of the Gulf's major prawn and pearl fisheries.[112] Thousands of green turtles faced a catastrophic breeding failure as their nesting sites were cut off by the slick.[113] Although there were hardly any attempts to protect the environment from the oil spills, the International Maritime Organization spent $4 million of its own money to clear the prime turtle nesting areas, so very few turtles died.[114]

The May 1991 issue of *Scientific American* reported that the US Department of Energy had released a memorandum to its numerous facilities saying this:

> DOE Headquarters Public Affairs has requested that all DOE facilities and contractors immediately discontinue any further discussion of war-related research and issues with the media until further notice. The extent of what we are authorized to say about environmental impacts of fires/oil spills in the Middle East follows: "Most independent studies and experts suggest that the catastrophic predictions in some recent news reports are exaggerated. We are currently reviewing the matter, but these predictions remain speculative and do not warrant any further comment at this time."[115]

This directive, it seems, still stood some months after the end of the war, when blazing oil wells continued to spew over 50,000 tons of sulfur dioxide (an acid rain producer) and 100,000 tons of soot a day.[116] The world must be told about the environmental consequences of the Persian Gulf War; the Department of Energy has no right to censor this devastating information for its own protection. Of particular importance, George H. W. Bush founded Zapata Oil; former Secretary of State James Baker is linked with Exxon, Mobil, Standard Oil, and Kerr-McGee; former Vice President Dan Quayle is backed by Standard Oil; and the Department of Defense is the world's largest consumer of oil.[117]

Our addiction to oil is once again threatening the very foundation of the global food chain and life-support system. The addiction is so strong that we must fight for oil, spill it, and, in the process, destroy even more of the planet. It took one million years to produce the quantity of fossil fuel that the global economy consumes each year.[118]

Nuclear meltdowns, poisoned gas, over a hundred thousand human deaths, dead birds, greenhouse gases, acid rain, toxic byproducts from oil fires, and new oil spills—all for oil. What will we do next? As I answer this question in the discussion that follows, I am reminded of an alcoholic fighting for his next drink, smashing family treasures, hurting his children, killing his wife, and burning down his house.

Which brings us to the second Gulf War, the shock-and-awe invasion of Iraq orchestrated by George W. Bush and his neocon mates—Donald Rumsfeld, Dick Cheney, and others. Following 9/11, officials at the highest levels in the White House decided that the terrorist attack on the United States provided a perfect opportunity to invade Iraq again, even though the perpetrators of 9/11 had no relationship with that nation. It took some time to organize the invasion because there was a spontaneous worldwide protest and outpouring of dissent, but in the end the Bush administration prevailed. The shock-and-awe operation began in March 2003 and has been an absolute catastrophe. Over one million Iraqi civilians have been killed,[119] more than three million people have been rendered homeless and turned into refugees within their own land or forced to flee to neighboring countries, and the whole infrastructure of the country—sewage, electricity, hospitals, transportation systems, schools, museums, archaeological sites—has been effectively destroyed.

Specific ecological damage has occurred because the United States and the "coalition of the willing" have been using depleted-uranium antitank weapons that burn upon impact, producing tiny aerosolized particles that can be inhaled into the lungs and that will be blown about in the desert winds for the rest of time, contaminating water, food chains, and lungs. The half-life of this uranium-238 is 4.5 billion years, and over 2,000 tons of this ordinance has been used thus far in the second Iraqi invasion. This material is carcinogenic, and children are particularly vulnerable to its effects. In addition, unrestrained looting took place in 2003 at some of Iraq's one thousand abandoned nuclear facilities, while as many as four hundred looters per day sacked the Al-Tuwaitha 120-acre (about 50-hectare) nuclear complex south of Baghdad, stole nuclear material, and emptied buckets containing uranium to use for domestic purposes. Drums containing deadly thorium, cobalt, cesium, and plutonium were also opened. Desperate people lived in these nuclear facilities while children bathed in the radioactive water.[120] Water, electricity, and sewage systems were severely damaged in many towns and cities, leading to an increased incidence of waterborne diseases, including cholera and typhoid, particularly among children.

Now the ultimate has been achieved. More than four thousand US troops had been killed, and at least fifty thousand had been injured by March 2007.[121] The war, it is estimated, will cost at least $3 trillion dollars, according to economist and Nobel Laureate Joseph Stiglitz. After forty years in the wilderness, evicted by Saddam Hussein when he nationalized Iraqi oil so that it belonged to the people, the major foreign oil companies have returned to reclaim Iraqi oil, the third largest deposit in the world, so that we can continue driving our cars and SUVs, ensuring that irreversible global warming will occur and that our children and grandchildren will have a grisly future.

Sewage

In my medical school days, we had to inspect the sewage "farm" in Adelaide, Australia, as part of our public health course. I remember feeling somewhat nauseated as I watched a man shoveling feces from a trench into a holding vat. When this material matured, it was used to fertilize fields and to grow food, after the proper elimination of pathogenic bacteria and parasites. China has for centuries been using human excreta as fertilizer. Traditionally, sewage has been a prized asset in China, and people who control the town sewage system have been among society's elite. When I visited China in 1988, I was amazed by the extraordinary abundance of the crops. Human waste is actually one of the most effective forms of fertilizer, and I have always wondered why we throw it away instead of recycling it. It comes from our food; it should go back to the food chain.

Today's sewage, however, is polluted with toxic household chemicals and industrial waste. Apart from clean human waste and soap, nasties such as cleaners, bleaches, paints, heavy metals, pesticides, arsenates, cyanide, medical drugs, and radioactive isotopes join the waste stream. These chemicals obviously make sewage extremely dangerous. In fact, a rather obvious side effect of modern-day life is the preponderance of an array of pharmaceuticals, including anticonvulsants, hormones, anti-inflammatories, and mood stabilizers, in the drinking water of forty-one million

Americans. Passed in urine and feces, these drugs are not filtered out in the waste treatment process, but the processed wastewater is nevertheless discharged into lakes, rivers, and reservoirs. In Philadelphia in 2008, fifty-six drugs were found in the drinking water, and sixty-three pharmaceuticals or pharmaceutical by-products were detected in the city's watersheds. These drugs are found in small concentrations, but they could well be influencing the physiology of millions of unsuspecting people.[122]

If sewage is disgorged into the sea, many of these chemicals concentrate in fish. Sydney, with a population of 4.1 million people, has ten coastal treatment facilities, but raw sewage contaminated with bacteria and chemicals is discharged into the ocean quite close to some of the most beautiful fishing and surfing beaches in the world. Roughly 0.5 percent of Sydney's sewage flows into the ocean untreated.[123] Sydney is a wonderful city—more beautiful, some say, than San Francisco—yet the state government is so shortsighted that it has blighted tourism by contaminating these beaches with raw sewage. Syringes, condoms, tampon applicators, plastic bags, and feces wash up on the golden sands.

The Mediterranean has been similarly contaminated. In 1990, several hundred dead dolphins washed up on the beaches of Majorca, the Costa Brava, and other Spanish resorts. Veterinarians concluded that toxic chemicals in the water had weakened the dolphins' immune systems. In August 2007, dozens of animals, especially striped dolphins, were infected and often killed by a virus. Earlier in the summer, hundreds of jellyfish had plagued Mediterranean beaches. Marine experts believe that these incidents mark the beginning of an epidemic related to warmer sea temperatures and polluted waters.[124]

The azure-blue waters of the Mediterranean have been terribly polluted over the last few decades by the enormous industrialization of southern Europe and by the fifty million tourists on the beaches. The World Wildlife Fund predicts that between 235 million and 355 million people a year will visit Mediterranean coasts by the year 2025.[125] The Rhône River alone, which flows past some of France's most potent petrochemical and nuclear power plants, daily deposits into the sea an enormous amount of nitrates, sulfates, iron, sewage, mercury, and pesticides.[126] And people in New York and New Jersey were dismayed some years ago to find their beaches covered with garbage similar to that found on Sydney beaches. Pollution has become universal.

Sewage contains large concentrations of phosphorus, an element used in detergents to soften water. Phosphates are potent plant fertilizers. When phosphate-rich sewage enters water systems, algal growth is promoted, causing many waterways to become clogged with algal blooms and overgrowth. The algae may totally cover the surface and, as they grow, they absorb the dissolved oxygen in the water. The subsequent oxygen depletion causes fish kills. The phenomenon is called "eutrophication." A perfect example of eutrophication was exposed to world view in 2008 when the sea around Qingdao that was to be used for rowing at the Beijing Olympic Games became totally clogged with tons and tons of bright green algae caused almost certainly by the discharge of nutrient-rich, untreated sewage—runoff from fertilizer and industrial pollution.[127]

Radioactive Waste

Oceans are also contaminated with radioactive waste. Sellafield, a large nuclear reactor complex in Britain, has expelled huge quantities of radioactive elements into the Irish Sea, making it the most radioactive sea in the world. The elements include cesium-137, strontium-90, and plutonium-239. When released into the environment, these and many other radioisotopes concentrate thousands of times at each step of the food chain—from algae to crustaceans, to small fish, to big fish. It is almost certainly dangerous to eat fish caught in the Irish Sea, and the danger may extend great distances. Not all fish are radioactive, but fish swim thousands of miles, and it is impossible to detect which fish carry radioactivity in their flesh. When humans eat radioactive food, the elements are absorbed through the wall of the gut into the bloodstream, to be deposited in specific organs. Strontium-90, which remains poisonous for six centuries, deposits in bone; cesium-137 (equally long-lived) settles in muscle; and plutonium-239 concentrates in bone, liver, spleen, and testicles and crosses the placenta into the developing fetus.

Although these three elements are all potent carcinogens, plutonium is the most dangerous. Plutonium remains radioactive and poisonous for half a million years, and just one pound (about half a kilogram) of plutonium could induce lung cancer in every human lung on Earth if adequately distributed. For example, if I die of a plutonium-induced cancer and am cremated, the smoke and plutonium from my body will leave the crematorium chimney and be inhaled by another human being; thus the cancer cycle continues ad infinitum. (The latent period of carcinogenesis is five to fifty years—the period of time before cancer develops after exposure to radiation.)

By now, over 1 million gallons (3.8 million liters) of radioactive waste containing, among other elements, about 550–1,100 pounds (250–500 kilograms) of plutonium has entered the Irish Sea.[128] And wherever nuclear power plants are located, radioactive waste is discharged into seas, rivers, or lakes. All reactors need millions of gallons of water a day for cooling, and this water is routinely flushed back into the water system, inevitably polluted with radioactive elements.

Another source of radioactive pollution of oceans is sunken nuclear submarines. Many of these ships carry not only nuclear reactors but also nuclear weapons, each containing 10 pounds (4.5 kilograms) of plutonium. As the reactor and bomb containers rust, radioactive elements inevitably escape into the sea. Since 1945, twenty-seven submarines have sunk, including five Soviet, four US, three British, and three French vessels.[129] There are at least fifty nuclear bombs on the ocean floor and eleven nuclear reactors that once powered submarines, and there remain approximately twenty-five thousand to thirty thousand nuclear weapons in the world's arsenals.[130]

In the early 1990s, the United States and Japan reached an agreement allowing Japan to reprocess its spent or used civilian nuclear fuel in France. Four hundred pounds of plutonium have been shipped by air and sea back to Japan over a period of thirty years.[131] It should go without saying that the chances that a serious accident—one in which the highly lethal plutonium could be released—will occur during such transport are high.

In 1999, a nuclear accident occurred at Japan Nuclear Fuel Conversion Company (JCO), a subsidiary of Sumitomo Metal Mining Company in Tokaimura, Japan. Workers added enriched uranium to a precipitation tank, which sparked a nuclear chain reaction. The International Atomic Energy Agency (IAEA) concluded that it was an "irradiation" accident, not a "contamination" accident, because not a significant amount of radioactive material was released. What the IAEA failed to explain was that the uranium had reached critical mass and huge quantities of neutrons had been released. Neutrons travel through concrete, steel, and plaster; and they travel distances of many miles. They are extremely mutagenic and carcinogenic; hence, many people in the village of Tokaimura would have been exposed, meaning that later in their lives they could develop cancer. Three JCO workers were hospitalized, and two died several months later, having absorbed a lethal dose of radiation.[132]

Now let's move back to land to examine radioactive pollution created by the nuclear weapons industry over the last sixty-five years. During the cold war, the US Department of Energy (DOE) supervised the construction of more than seventy thousand atomic and hydrogen bombs, contracting the work out to private corporations, including Rockwell International, Dow Chemical, General Electric, Westinghouse, Kerr-McGee, and DuPont. The building of seventeen huge nuclear facilities around the country proceeded in total secrecy in order to protect "national security." Hence, neither the DOE nor the private contractors were held accountable for nuclear accidents, radioactive spills, nuclear meltdowns, leaks, and deliberate releases into air, water, and soil.[133]

Not only were tens of thousands of workers contaminated over time, often without their knowledge, but hundreds of thousands of US citizens were irradiated. Some populations were actually used as guinea pigs and exposed to radiation as part of an experiment. Many people are now developing cancer, and their babies are being born deformed. Animals living near the nuclear weapons plants were also exposed to clouds of radioactive isotopes, and there have been clearly defined epidemics of deformities directly related in time to the release of radioactive materials. Radiation, in levels exceeding those at Chernobyl, was systematically released into the air and water at the Hanford bomb facility, in Washington State, more than forty years ago.[134]

A series of articles in the *New York Times* exposed these acts of negligence when the information first became available for public scrutiny, in 1988 and 1989, and I will quote from several of these stories. Remember that carcinogenic chemicals are biologically synergistic with radioactive isotopes in human and animal bodies.

In Hanford, Washington, a complex of nuclear reactors and 177 underground tanks containing high-level radioactive waste is located adjacent to the Columbia River. The tank complex is euphemistically called a "tank farm." The waste material contains over a hundred long-lived isotopes, including plutonium, cesium, and strontium, dissolved in concentrated nitric acid. These must be isolated from the environment for thousands of years.[135]

Not only is nitric acid extremely corrosive, but the radioactivity causes the liquid to boil at high temperatures. Amazingly, the DOE kept no record of the precise content and concentration of isotopes in the tanks. In the mid-1950s, potassium ferrocyanide was added to the tanks in order to concentrate the radioactive waste.[136] More recently, experts recognized that cyanide combines with nitric acid to form an explosive, combustible mixture, and that it is therefore possible that the tanks will explode, scattering the deadly contents much as the Chernobyl explosion did. Many of the tanks have a single-walled steel construction, and between 1958 and 1975 twenty of these tanks developed cracks and 430,000 gallons (about 1.5 million liters) of nuclear waste material escaped. This material is migrating toward the Columbia River.[137] In all, 700,000–900,000 gallons (2.6 million–3.4 million liters) have leaked.[138] Twenty-seven streams around Hanford are contaminated with radiation, and rabbits and coyotes using the radioactive waste as salt licks have scattered radioactive dung over an area of 2,000 acres (about 800 hectares).[139]

Until 1971, the nuclear reactors were cooled with water from the Columbia River, and the radioactive coolant was discharged back into the river on a daily basis. From 1945 to 1971, tens of millions of curies (a curie is a measurement of radiation) were dumped into the river. In the early 1960s, a Hanford worker ate oysters caught hundreds of miles downstream at the mouth of the Columbia River. When he went to work the next day, he set off the radiation alarm at the Hanford plant.[140] The Columbia River is a rich spawning area for salmon, and I have seen fishermen in Astoria at the mouth of the river catching salmon. Why are they not warned about the dangers, and how many thousands of other people, not checked by radiation monitors, are unknowingly radioactive?

In September 1991, the General Accounting Office announced an investigation into the mysterious disappearance in 1989 of key documents that calculated leaks from the radioactive waste tanks. Apparently, 750,000 gallons (about 2.8 million liters) of cooling water was pumped into a leaking tank, and 750,000 gallons of radioactive liquid escaped. The GAO also reported that 444 billion gallons (about 1.7 trillion liters) of liquid emanating from reactors, processing plants, and tanks was discharged into the environment at Hanford between 1945 and 1991.[141]

In July 2007, workers at the Hanford Nuclear Reservation site tried to unclog a pump with high-level waste in it. In doing so, they accidentally forced the waste from an underground tank into a water line that leaked. An estimated 50–100 gallons (about 190–380 liters) of radioactive waste spilled into the ground.[142] Such occurrences are not uncommon at Hanford.

Not only has the Hanford plant been discharging and leaking radiation into the Columbia River for forty-five years, but serious accidents have occurred at the reactors. In addition, on several occasions huge clouds of isotopes were created knowingly and willfully. In December 1942, about 7,800 curies of radioactive iodine-131 was deliberately released in an experiment designed to detect military reactors in the Soviet Union (by comparison, only 15–24 curies of iodine-131 escaped at Three Mile Island in 1979). Then, between 1944 and 1955, over half a million curies of iodine-131 was released.

Radioactive iodine concentrates in the food chain, particularly in cow's and human milk. Babies and young children are ten to twenty times as sensitive as adults, because duplicating genes in actively dividing growing cells are particularly vulnerable to damage from ionizing radiation; and iodine specifically concentrates in the thyroid gland. An abnormally high incidence of thyroid tumors and cancers has been observed in populations living downwind from Hanford. Strontium-90, cesium-137, and plutonium-239 have also been released in large quantities, as was ruthenium-106 between 1952 and 1967. People in adjacent neighborhoods were kept uninformed about these releases before, during, and after—and none of them were warned that they were at risk for subsequent development of cancer.[143] Some experts estimate that downwind farms and families received radiation doses ten times higher than those that reached the Soviet people living near Chernobyl.[144]

In all, over 80 square miles (207 square kilometers) at Hanford are contaminated with deadly, very long-lived radioactive elements, including iodine-129, strontium-90, technetium-90, tritium, and uranium, as well as chromium, which is also carcinogenic, and the solvent carbon tetrachloride, which will help the migration of these elements. The DOE calculates that 450 billion gallons (about 1.7 trillion liters) of radioactive waste has been "released" into the ground, some of which is approaching or has entered the Columbia River. Radioactive animals abound on the reservation, including gophers, wasps, spiders, snakes, birds, and fruit flies.[145]

So what does the DOE do? It organizes sightseeing tours complete with cheery PhD holders explaining the history and wonders of this Hades-like establishment that epitomizes the madness of the nuclear age.

A thirty-seven-year-old uranium milling plant at Fernald, Ohio, near Cincinnati, has, since 1951, released 298,000 pounds (135,000 kilograms) of powdered uranium into the air, 167,000 pounds (76,000 kilograms) into the Great Miami River, and another 12.7 million pounds (5.8 million kilograms) into earthen pits.[146] Uranium, the fuel for nuclear reactors, is also used in casings for hydrogen bombs. When fissioned in a nuclear reactor, it is converted into more than two hundred isotopes, all biologically more dangerous than the original uranium. It is nevertheless a potent carcinogen in its own right—one that, if swallowed, migrates from gut to bone. Several years ago, a young boy living near the Fernald plant, developed osteogenic sarcoma (bone cancer). After his leg was amputated, the bone was analyzed and found to be contaminated with high concentrations of uranium. Incidentally, the uranium milling plant was for many years disguised from the local population; it was called the "feed materials production center," masquerading as a pet food factory.

Other radioactive weapons plants are located at Oak Ridge, Tennessee; at Rocky Flats, Colorado; on the Savannah River in South Carolina; at West Valley, New York; and at Idaho Falls. The DOE admitted for the first time in the 1990s that hundreds of thousands of people living around these facilities have been exposed to accidental "releases" since 1945. Similar stories of tragedy and grief are common among all these populations.[147]

An unusual cluster of primary brain cancers in Los Alamos spurred the New Mexico Department of Health and the US Department of Energy to initiate a study to determine how much radiation the Los Alamos National Laboratory has released into the environment since 1943 and to assess whether the incidence of brain cancer in the community of eighteen thousand people is excessive. Over the years, the laboratory released millions of gallons of radioactive and toxic wastes, which were dumped secretly into the nearby canyons and ravines. But the DOE has no official documentation regarding the quantities dumped or even the location of the dump sites.[148]

In the 1960s, targeteers at the Pentagon told the DOE to cease manufacture of bombs because they had all relevant targets in the Soviet Union covered, but bomb production continued, as if by some sort of gigantic, autonomous, uncontrollable production-line monster. At the end of the cold war, the DOE was still manufacturing up to five new hydrogen bombs every day.

In 1989, Congress appropriated $240 billion to "clean up" fifteen of these weapons-producing sites. The problem is the same as at Love Canal: how does one retrieve radioactive elements long since scattered to the four winds, hiding in dead and living bodies (animal and human), migrating from earthen pits, leaching into underground aquifers and rivers, and concentrating forevermore in radioactive food chains?[149] C'est impossible! Apart from the radioactive waste at these weapons sites, there are nearly four thousand poisonous solid-waste sites—no one knows exactly how many because record keeping by the DOE and its contractors over the years has been bad.[150] Although it had a thirty-year plan to "clean up" these radioactive sites, the DOE admitted that it is a next-to-impossible task—when you dig up the waste, where do you put it?[151]

Catastrophes at nuclear weapons plants reflect similar accidents at civilian nuclear reactors—some leaking radiation, and all creating massive quantities of radioactive waste on a daily basis. There are 104 active plants in the United States, and 438 reactors worldwide.[152] The Nuclear Regulatory Commission estimated that there was a 45 percent chance that a meltdown larger than Chernobyl's would occur in the United States between the years of 1985 and 2005.[153] Although such a disaster has not yet happened, it is not a matter of *if*, but *when*. Because radiation knows no boundaries, the consequences of a meltdown affect many countries. Fallout from Chernobyl spread not just to the Soviet Union but also to large areas of Europe, extending through Poland, Austria, Germany, northern Italy, France, the Scandinavian countries, and parts of Britain. The decimated Chernobyl reactor with its intensely radioactive fuel burned for ten days, spewing four hundred Hiroshima bombs' worth of radioactivity across Europe.[154]

An article in 2007 predicted that 800,000 children in Belarus and 380,000 in the Ukraine[155] are at risk for developing leukemia in the wake of the Chernobyl accident. It also reported that many babies in the region have been born without arms or legs or with other gross deformities. Between 1990 and 1994, UNICEF reported that nervous system disorders increased by 43 percent, cardiovascular diseases by 43 percent, bone and muscle disorders by 62 percent, and diabetes by 28 percent.[156] In 1991, the Ukrainian Ministry of Health reported an increase in the number of miscarriages, premature births, and stillbirths, and three times the normal rate of deformities. The World Health Organization predicts that fifty thousand children in the Gomel Region of Belarus will develop thyroid cancer during their lifetime because they drank contaminated milk and inhaled radioactive iodine at the time of and in the weeks following the accident.[157] Up to the year 2005 in the Ukraine alone, 2.3 million people (including 452,000 children), were treated for cancers related to radiation, such as thyroid and blood cancers.[158] Russian scientists call the radiation sickness, and associated cancer and leukemia, "Chernobyl AIDS" because the disease bears a close medical resemblance to AIDS. Since the accident, the incidence of heart disease has tripled in Belarus, and doctors have reported a high incidence of numerous defects of the heart, now known as "Chernobyl Heart."[159]

Over 21 percent of Belarusian farmland remains contaminated. Livestock such as cattle and goats continue to accumulate radiation in their meat and milk, and this radiation finds its way into humans. In Belarus alone, 2,697 villages, with a combined population of two million people, have been seriously contaminated. People in Russia are still eating radioactive food (food grown in contaminated areas will be radioactive for thousands of years). Food and water are still contaminated. Mushrooms, in particular, are highly contaminated in Belarus, but they are still collected and eaten. The land and food will not be safe again for five hundred thousand years because that is the amount of time that plutonium, the most deadly of radioactive elements, remains dangerous.

On the fifth anniversary of the Chernobyl disaster, I was interviewed on the radio along with Robert Gale, the doctor who had treated many of the acute radiation victims, and John Gofman, a world-famous expert in radiation. It was clear at that time that Chernobyl would continue to be a tragedy, and doctors were still not sure of the true medical consequences. Leukemia takes five to ten years to become clinically apparent, and solid cancers, fifteen to sixty years, but the medical consequences are now frighteningly apparent.

Despite this uncertainty, new data in 1991 revealed that approximately 650,000 people participated in the cleanup operation after the accident and that these people, called "liquidators," were exposed to very high levels of radiation. According to various estimates, between five thousand and ten thousand of them have already died from radiation-induced illnesses. Five million people (one-quarter of them children) still lived in contaminated areas five years after the accident; and three hundred thousand, in highly contaminated places.[160] An estimated 3.1 million people are living in less contaminated areas, such as Belarus, parts of Russia, and the Ukraine.[161] The Russian government does not have the facilities or the resources to relocate these people or adequate medical equipment to monitor the internal radiation levels, let alone to provide bone marrow transplants and medical treatment for the many expected leukemia victims of the future.

What Must Be Done

Although there are no definitive scientific answers to the most serious problem of radioactive waste, certain issues must be addressed immediately:

- All nuclear reactors, both military and civilian, must be closed down at once so that no more waste is produced, no more bombs are made, and no more accidents occur.
- More than one-third of all the engineers and scientists in the United States work for the military-industrial complex.[162] Since the cold war is over and war is clearly obsolete, these scientific brains should be transferred from weapons manufacture to the urgent task of finding safe radioactive-waste storage.

We did not inherit the Earth from our ancestors; we borrowed it from our descendants. If we fail to solve this dreadful problem, radioactive food and water will be our legacy to future generations, with increased incidences of genetic disease, deformed babies, and epidemics of children dying of cancer and leukemia.

With all of this evidence before us, we must now examine the people running for president and our most influential politicians. Unfortunately, almost all of them are in favor of nuclear power to counter global warming. They are clearly ignorant, and they need to be well and truly educated by their constituents—that's you!

Apart from the alarming medical problems outlined in this chapter, nuclear power in fact adds substantially to global warming in its own right because of the huge amounts of fossil fuel used to mine and develop the uranium fuel, to construct the reactor, and to transport and store the radioactive waste for half a million years.[163] House Speaker Nancy Pelosi once opposed nuclear power, but now she is for it. Senators John McCain and Barack Obama both favored nuclear power. Now President Obama has an ambivalent attitude toward nuclear power.

We have a lot of work to do!

6 Species Extinction

In my book, a "pioneer" is a man who turned all the grass upside down, strung barbed wire over the dust that was left, poisoned the water and cut down the trees, killed the Indian who owned the land and called it progress. If I had my way, the land would be like God made it, and none of you sons of bitches would be here at all.

— Charlie Russell, cowboy artist, 1923

We have taken over the planet as if we owned it, and we call it progress because we think we are making it better, but in fact we are regressing. Species are dying in the wake of this "progress," and we seem not to realize that our life depends on theirs. Peter Raven, director of the Missouri Botanical Garden in St. Louis, says that the destruction of species is more critical for the world than are the greenhouse effect and ozone depletion, because it is moving faster and is inevitable. Up to 270 species become extinct every day.[1] For fifteen years, I traveled the world warning people about the medical and ecological consequences of nuclear war, and I was not aware that life was already dying quietly and unobtrusively as a result of humans' ongoing activities. Now I see that the threat of species extinction is as serious as the threat of nuclear war.

Life began on the planet four billion years ago, and over time an astounding array of diverse forms has gradually emerged. But the advance of evolution has not always been smooth. The number of species increased rapidly up to 600 million years ago, and then decreased for the next 200 million years. The last 400 million years have been characterized by slow increases in numbers, interrupted by five significant extinction phases. The largest of these was during the Permian era, 240 million years ago, when 77–96 percent of all marine animal species became extinct. It took another five million years for species diversity to recover. About sixty-five million years ago, another significant era of species extinction began when the dinosaurs, which had previously ruled the Earth, disappeared, and mammals gained global hegemony. Thus, the evolutionary stage was set for the appearance of *Homo sapiens*. Since that time, the number of species has continued to increase to the present, all-time high.[2]

Human beings first appeared in a primitive form some three million years ago. The species lived in relative harmony with other life-forms until ten thousand years ago, when it began to have a devastating effect on the diversity of other species.

Humans began hunting animals and birds and chopping down or burning forests and plants. To give two examples, in Polynesia, an isolated island environment, half of the bird species are now extinct because of human hunting and forest destruction; and more recently, during the 1800s, almost all the unique shrubs and trees were destroyed on the small island of St. Helena, in the North Atlantic. Of the ten million to thirty million species estimated to be extant today, we may now be losing fifty thousand each year, according to Professor David Tilman of the University of Minnesota.[3] Current extinction rates are between a thousand and ten thousand higher than background rates—the rates at which some species naturally became extinct during the process of evolution.[4] Some twenty thousand species are known to be threatened with extinction[5]—one in three amphibians, one in four mammals, one in eight birds, and one in four coniferous trees.[6]

In this time frame, the span of existence of *Homo sapiens* seems trivial. Yet we now threaten to exterminate most of the world's species, which took four billion years to evolve to this point. Although the dinosaurs disappeared, they did not, as we may well be doing, bring about their own extinction.

Unlike the dinosaurs, we clearly have almost total dominion and control over the planet. The development of the opposing thumb gave us the ability to make and hold instruments, weapons, and tools of mass destruction. Our abnormally large neocortex—the thinking part of the brain, which developed in a very short evolutionary time frame—has enabled us to communicate thoughts by speech and writing and to destroy, dominate, and subdue all other species for our own benefit, as we perceive it. We are shortsighted and egocentric, hardly realizing that our survival is intimately related to and dependent on the survival of up to thirty million other species.

This reckless and irresponsible behavior may, in the long run, be a kind of suicide. The behavior of most other species is conditioned for their long-term survival, but humans, because of the rapid evolutionary development of the intellectual neocortex, are able to control and dominate nature, and in so doing they are destroying the foundation of their biological existence. Humans stand at the apex of the pyramid of the food chain. Bacteria in the soil break down the fallen leaves to produce humus and compost to feed the plants to feed us. If we kill the bacteria with chemical toxins and change of climate, we will indirectly kill ourselves. From another perspective, if we upset the balance of nature by eliminating large predatory animals, we produce a reactionary overpopulation of smaller predators that are normally kept in check. These, in turn, then eat and render extinct the lower-order animals upon which they feed. We therefore must not disturb the hierarchical balance of nature and the food chain.

We have a similar relationship with many other species.

We really "came into our own" with the dawn of the industrial age, early in the 1800s. As we harnessed nature and worked with the natural laws of science, we learned to destroy forests and pollute the air, water, and soil very efficiently, and this efficiency has, over the last two centuries, increased exponentially. In the grand scheme of evolution, our obsession with interpersonal, national, and religious conflicts, and our wars are meaningless. Unfortunately science has been used over the last century to make killing more efficient, so much so that now we can destroy almost all planetary life with the press of a button.

Although five million to thirty million species of plants and animals have not been systematically documented, scientists arrive at this number by extrapolation, having analyzed the number of new species in a small area of rain forest. The International Union for Conservation of Nature (IUCN) recorded and named 1.5 million species, but scientists believe that a hundred million species have yet to be named. Within each of these documented five million to thirty million species, there is a huge degree of genetic diversity that is terribly important. Let me use the species of *Homo sapiens* as an example. Each person carries a unique set of genes that was derived from a particular sperm that fertilized a particular egg. Each sperm and egg are genetically different from any other sperm or egg, and this unique gene structure (genotype) determines all the facial and bodily characteristics, personality, and mental abilities of an individual. The process of evolution depends on genetic diversity, because as the environment changes, only those organisms with specific characteristics that allow them to survive the change will reproduce. This is what Charles Darwin called "survival of the fittest."

Without genetic diversity, evolution could not have happened. To give a simple example, some breeds of maize are able to survive better in climatically unfavorable conditions than are others. Scientists are finding that by mating one variety of wild maize with a domestic type, a better crop plant can be produced. So instead of concentrating on saving a few individuals within a particular species, we must save all the variants of life-forms within each species. But instead, we look on, as between 2.7 and 270 species become extinct every day.[7]

There are thousands or even millions of varieties of plants that we have not yet even identified, but as we destroy the environment, we will be needing special strains of wild maize, wheat, rice, and other species that will grow in difficult terrains and climates. Over thousands of years, the human race has utilized about seven thousand different plant species for food, but the present generation tends to rely on only about twenty species to provide 80 percent of the world's food. These twenty include rice, wheat, millet, and maize.[8]

We consume less than 0.1 percent of naturally occurring species.

Yet we know that more than seventy-five thousand plant species are edible and that some are far more appropriate for consumption than are those we now use. Edward O. Wilson described a plant called the winged bean, or *Psophocarpus tetragonolobus*, from New Guinea, which is at present ignored by the world's food manufacturers and farmers. The whole plant is edible—roots, seeds, flowers, stems, and leaves—and a coffeelike drink can be made from the juice. It grows rapidly, to a height of 15 feet (4.6 meters) in several weeks, and exhibits a nutritional value equivalent to that of the soybean.[9] At a time when the human population is growing explosively and needs an enormous amount of food, it seems imperative that we start preserving and cultivating different plant and food varieties that will provide efficient sources of nutrition.

Ironically, as other species become extinct, we are proliferating. In 1800, we numbered 1 billion; in 1990, 5.2 billion. In the year 2008, there were 6.72 billion people on the planet. By the end of this century, if present trends continue, we may reach 14 billion—more than twice our present number.[10] Clearly, the ecosphere cannot sustain 6.72 billion, let alone 14 billion. (Overpopulation is discussed in Chapter 7.)

Let's look at the variety of ways in which species are being destroyed.

Tropical Forests

As noted in Chapter 4, tropical forests probably contain fifteen million to twenty-four million of the thirty million planetary species. Although Brazil has environmental laws to protect the Amazon rain forest, there is little government enforcement of the laws and 80–90 percent of the farmers ignore them. The latest threat to the forest is soy farming. When market prices for soy soared in 2004, the second worst deforestation in history took place. At least 17 percent of the Amazon forest is already gone, and if the current trend continues, 40 percent will be destroyed by 2050.[11] Within fifty years, the trees and the species inhabiting them may well be gone, destroyed because of Third World debt and First World greed.

Wetlands

Wetlands serve as crucial breeding grounds for fish, crustaceans, and other sea-dwelling creatures. Until very recently, though, mangrove swamps and reedy wetlands were not valued as habitats for species. Rather, they were seen as ugly, muddy, difficult areas that were best cleared for canal developments or filled in for real-estate investment. Although some people now understand the ecological significance of wetlands, most developers are still uneducated. Indeed, the majority of people in the Western world are deeply ignorant about the biological meaning of species extinction.

About 50 percent of the global wetlands have so far been destroyed. In the United States, the more agricultural states have lost over 80 percent of their historical wetlands, and there are only tiny and highly disturbed remnants in cities such as Los Angeles and New York.[12] Studies of avian communities show that re-created wetlands in forested areas (re-creations required by the US government if anyone—for example, a logging outfit—destroys the habitats) lack the richness and diversity of natural wetlands and harbor a different composition of avian species. Thus, re-created forest wetlands cannot provide a full ecosystem replacement for natural wetlands.[13] By the 2080s, 30 percent of coastal global wetlands will be gone, along with the fish that once inhabited these marshes and served as a key source of protein for millions of people.[14]

A potent example of wetland destruction is occurring at Dalai Lake and its surrounding wetlands in China, which are of "international importance" because they stand at the intersection of the Chinese, Mongolian, and Russian borders. In this crucial staging ground for the annual migration of almost three thousand bird species, China's largest gold-mining company, China National Gold Group, is secretly laying a huge pipeline to drain water from the lake to be used to leach the gold out of the mine's low-grade ore. This operation will hasten the already rapid drop in the water level of the lake and dry out much of the wetlands, threatening the very existence of many of these species. Furthermore, in its infinite wisdom, the World Bank has funded a freeway through the marshland, destroying much of the birds' habitat.[15]

Frogs: The New Canaries

Years ago, canaries were used as sensitive indicators in coal mines to determine whether the air was safe for the miners to breathe. When the canary died from toxic fumes, it was time to leave. For the last decade, biologists have noticed an alarming decline in the numbers and species of amphibians—frogs, toads, and salamanders. These creatures were the first vertebrates ever to inhabit the land. They appeared four hundred million years ago and evolved into present-day species some two hundred million years ago. Because their bodies are acutely sensitive to the environment, they are our canaries—our barometer of global environmental poisoning.

Frog species are disappearing from Australia, the United States, Japan, Canada, Puerto Rico, Costa Rica, Brazil, China, India, and Peru. According to the latest IUCN "Red List of Threatened Species," one in three amphibian species is threatened with extinction. Not until 2006 were amphibians assessed completely and found to have a higher percentage of critically endangered species than either birds or mammals.[16] This phenomenon is particularly worrisome because species are disappearing from national parks and from some of the best-protected areas of the world, where they should be safe.

What is threatening them? Many theories have been proposed:

- Frogs' eggs are very sensitive to the increased UV light that is due to the thinning ozone layer.
- Foreign predator fish introduced into frog habitats, such as ponds and lakes, eat tadpoles and baby frogs.
- Toxic chemicals and heavy metals are poisoning the frogs because frogs naturally absorb large amounts of water, some of which is now polluted, through their skin.
- Habitat damage caused by logging, pesticide pollution, and dam construction threatens frogs' survival.

- Frogs are so sensitive to variations in temperature and moisture that if the rainfall and climate change, the frogs die—and these changes seem to be occurring as the greenhouse effect becomes manifest.

All of these factors can overlap to stress the frogs' immune systems, which may be what led to the latest and most destructive cause of the frogs' decline: a worldwide spread of the lethal chytrid fungus. And further studies have shown that this chytrid epidemic is driven by global warming, pollution from agricultural chemicals, or increased exposure to ultraviolet radiation. However, there appears to be a ray of hope in Queensland, Australia. In the late 1970s, six Queensland frog species became extinct because of the fungus, but a reappearance and increase in the Kroombit tinker frog has been observed recently at several sites in the forests of Kroombit Tops in Central Queensland. Possibly the frogs developed an immunity to the fungus, or the disease either disappeared or became less deadly.[17]

Frogs are an important link in the ecological chain. Tadpoles consume large quantities of algae, thereby keeping streams clean and flowing, and adult frogs eat enormous numbers of insects, including mosquitoes. The Everglades, in Florida, offer a good example of the ecological necessity of frogs. Scientists recorded in the early 1990s a 90 percent reduction in the number of wading birds, possibly caused by the demise of pig frogs, which provided the birds' food.[18] If frogs are the new canaries, the situation is very serious and we had better act fast.

Coral Reefs

Coral reefs are a rich source of species and are very complex ecosystems in their own right. The Great Barrier Reef in Australia is the largest living structure on Earth, visible from the moon.[19] An area of about 150,000 square miles (400,000 square kilometers) of coral reef is estimated to contain five hundred thousand species. Less than ½ percent of the ocean floor is composed of coral reefs, but more than 90 percent of sea life is dependent on them.[20] Many of these animals and fish engage in a kind of biological warfare with one another, and they have therefore evolved a large number of specific toxins that can be harnessed for medical treatment and possibly other beneficial uses.

Tragically, coral reefs are mysteriously dying all around the world. It seems that coral polyps (the living organisms that provide the vivid colors and create the solid coral structure as their protective habitat) are very sensitive to temperature change. The year 1998 was an El Niño year and the warmest year on record in the twentieth century. (Already this century, 2005 was warmer, and there wasn't even an El Niño that year.)[21] In 1998, 16 percent of the world's coral reefs were damaged or destroyed. Not only does climate change warm the seas, but the increased concentration of dissolved CO_2 makes the water more acidic, and this increased acidity slows the building of the calcium carbonate in coral skeletons.[22]

Nearly 60 percent of the Earth's coral reefs are in danger of being destroyed over the next thirty years.[23] Toxic chemicals derived from agriculture, industrial waste discharged into rivers, and urban runoff from houses (pesticides, cleaners, fertilizers, and sewage), plus coastal development, sedimentation, destructive fishing practices, and tourism, are also killing the coral.[24] These magnificent reefs, true wonders of nature, must be preserved at all costs, and no change in climate or human development should be allowed to threaten their existence.

Arid Zones or Deserts

Deserts are habitats to many fragile species of plants and animals. Australian deserts were once filled with tiny exotic marsupials that looked like delicate, forlorn, scaled-down kangaroos, as well as other unique and precious life-forms.[25] Ninety percent are now extinct or threatened with extinction.[26] These indigenous animals were decimated by the introduction of foreign feral animals. Foxes prey on small animals, as do wild cats and dogs, while grass-eating rabbits, pigs, goats, camels, buffalo, rodents, wild horses (called brumbies), and cattle have destroyed thousands of square miles of fragile plant ecosystems in Australia.[27] Worldwide, the number of endangered plants and animals of the deserts exceeds four thousand. According to the IUCN, unregulated hunting and habitat degradation are the leading causes of this serious problem.[28]

Introduced Species

In the preceding section I described the devastation wrought by introduced species that have no natural predators. But sometimes foreign animals are brought into a country specifically to prey upon a natural pest that needs to be eradicated. For instance, the South American cane toad was brought to Australia to eradicate a beetle that was destroying sugarcane crops in Queensland. Not only did it not kill the beetle, which lived out of reach on tall sugarcanes, but outside its original environment and without its natural predators, the cane toad has multiplied out of control and spread like a creeping plague over much of the east coast of Australia. I live on the east coast hundreds of miles south of the point of introduction, and at night my paths and gardens are alive with silent, hopping, slimy cane toads, which are brown, warty, and ugly. They not only look repulsive, but have poisonous glands located in the area behind their head. Animals that eat them, such as snakes, goannas, lizards, cats, or dogs, are poisoned and die. The cane toad plague is now so serious that many of Australia's indigenous reptiles and other creatures are threatened with extinction. In the mid-1980s the Australian government ran out of money to research diseases that may control the cane toad. Efforts to control the toad population now are mostly local and volunteer—hunting, killing, or freezing the toads to death.

Savannas and Grasslands

The wildlife of savannas and grasslands is also under threat, particularly in Africa, because human beings enjoy displaying exotic animal skins on their floors, hanging animal heads on their walls, grinding rhinoceros horns for aphrodisiacs, wearing shoes and belts made from crocodiles and snakes, and making jewelry and piano keys from elephant tusks.

In Africa in the 1930s, there were ten million elephants; now there are fewer than three-quarters of a million. As of 2007, their population ranged from 470,000 to 690,000.[29] Eighty thousand elephants are killed each year by poachers who hack off their tusks with chain saws and leave the huge carcasses to rot in the midday sun, while they make a living from their illegal bounty.[30] Between 60 and 70 percent of the ivory is exported to the Far East and sold on the black market.[31] Before the 1989 ban on ivory harvesting put into place by the Convention on International Trade in Endangered Species of Wild Fauna and Flora, seventy thousand elephants were poached every year. Although poaching of their meat and ivory is still a problem, African elephants are now equally threatened by loss of habitat. As the human population proliferates, an ever-increasing area of their habitat is lost or degraded. In addition, their increasing isolation will have an undesirable impact on their genetic diversity.[32]

Poachers are often poor, and because the world supports an unequal distribution of wealth, this is their one mode of survival—the destruction of nature. Lions, tigers, leopards, zebras, giraffes, and other wonderful exotic animals are all endangered. Soon, these animals will exist only in zoos, and eventually these few remaining specimens will become extinct. For instance, the wonderful diversity of life that once teemed in the Florida Everglades is now increasingly threatened by the encroachment of sugar growers and by the biofuels industry.[33]

The total population of other creatures is also diminishing. It is predicted that rising acid levels in the Southern Ocean will start to destroy sea life within the next thirty years, causing irreversible destruction of shell creatures—the first to be affected.[34] The ocean's larger creatures, blue whales, have decreased in number from two hundred thousand to somewhere between three thousand and five thousand; and humpback whales, from fifty thousand to twenty thousand. Nineteenth-century whaling records suggest there were one hundred thousand humpback whales at that time.[35] The number of southern white rhinoceroses in Zaire dwindled from four hundred in the 1970s to fifteen in 1984; however, the population has now increased to five thousand individuals.[36] (The northern white rhino is now extinct.)[37]

Human activities are also driving primates to the brink. Almost half of the world's 634 types of primates, from giant mountain gorillas of Central Africa to the tiny mouse lemurs of Madagascar, are threatened with extinction because of human activity, including deforestation, hunting for human consumption, illegal logging, and land conversion. Tragically, the violence in the Congo is threatening endangered mountain gorillas.[38] At the twenty-second congress of the International Primatological Society in August 2008, however, Russ Mittermeier said, "It is not too late for our close cousins, the primates, and what we have now is a challenge to turn this around."[39]

Big cats are "magnificent creatures that are endlessly fascinating and wonderful to spend time with," according to Luke Hunter, the executive director of the Panthera foundation in New York. But "there is also this overwhelming need to do the conservation." Of the thirty-six species of big cats in existence, all have suffered a huge decline in numbers in recent years. The reasons are many, including human encroachment upon their habitats, conflict between farmers and animals for domestic livestock, hunting, and global warming. Tigers are the most imperiled, down to only five thousand in the wild, subsisting in only 7 percent of their original historical range.[40]

The Gouldian finch is an exquisitely colored bird that was once found throughout the grassy woodlands of tropical Australia. Fifty years ago, flocks of thousands used to swarm through the air, but the swarming has stopped. Now flocks of only fifty or fewer are seen, because the finch was extensively trapped for the caged-bird industry and because its habitat was destroyed by the regular, intentional burning of the grasslands. The bird nests are vulnerable to destruction because they are located in termite and ant mounds close to the ground, or in the hollows of old gum trees. A recent survey found only two intact breeding colonies in the whole country, and one of these was threatened by a gold-mining company. The company planned to mine at the site of the breeding colony and to construct tailing dams close by. These dams were to hold cyanide-contaminated water, so when the birds drank their daily fill of water they would die. Is this the epitaph of the Gouldian finch?[41]

Antarctica and the Arctic

One of the last bastions of pristine wilderness is now under threat. The Antarctic is an ecosystem delicately balanced and teeming with life. No land plants grow on the icy waste, but the sea supports all the life-forms. The base of the food chain consists of single-celled plants called phytoplankton, which trap energy from the sun. Billions of tiny crustaceans known as krill eat the plankton, which in turn forms the food base of whales, penguins, crabeater seals, and large seabirds such as the petrel. Small fish and squid also eat the krill and, in turn, feed emperor penguins, albatrosses, large fish, seals, and sperm whales.[42]

Although the climate of Antarctica is not conducive to human habitation, we have nevertheless devised many ways to intrude on and damage this unique and fragile biosphere:

- Large-scale international commercial fishing is depleting the sea. However, this is not the only known large-scale resource exploitation of the Antarctic. The most threatened species in this area is the Patagonian toothfish, known as the Chilean sea bass in the United States. Because this fish is valuable, it is subject to illegal, unregulated, or unreported fishing. These fishermen tend to be unscrupulous, and albatrosses in particular are threatened as "bycatch."[43]
- In February 1989, an oil spill from an Argentinean tanker killed thousands of penguins, skuas, and their chicks.
- The tourist industry recently decided that the habitat of the South Pole is a nice place to take its customers. Visitors leave a trail of desolation as tons of garbage are dumped from tourist ships into the sea. Incidentally, most ship captains in the world still believe that the oceans are a universal sewage disposal system and act accordingly.
- For over sixty-five years, scientists established research stations in the Antarctic, but they did not build adequate sewage systems and they left hundreds of cans of toxic waste and garbage when they departed. How could they research the delicate web of life and then so insensitively threaten it, or were

they interested only in their research papers? (Actually, the scientists at the US and New Zealand stations at McMurdo Sound now carry out all their waste.)

- Ozone destruction is an extraordinary threat. Remember that, in winter, the ozone depletion over the South Pole has, at some latitudes, reached over 99 percent.[44] NASA and the National Oceanic and Atmospheric Administration reported that in 2006 they had recorded the largest and deepest ozone hole in history. But the meteorological and ozone-monitoring unit of the British Antarctic Survey reported that the hole in August 2007 was the largest ever recorded.[45] Remember, too, that life cannot exist without the ozone layer. Plankton is extremely sensitive to UV light. As it dies, so will the rest of the life cycle.
- In addition to being extraordinarily precious to the food chain of the world, the Antarctic houses 70 percent of the world's freshwater, although some of this water is at risk, because the huge Wilkins Ice Shelf that joins the Antarctic coast is disintegrating rapidly. In July 2008 it was connected by only a small bridge of ice to the Antarctic mainland.[46] The Ross Ice Shelf, which is the size of France, is also at risk. Should this melt completely, ocean levels could rise 16–20 feet (5–17 meters).[47]
- Mining corporations have been pressuring their governments for the right to dig up minerals in the Antarctic, which would lead to the ecological devastation of much of the area. The Protocol on Environmental Protection to the Antarctic Treaty, known as the Madrid Protocol, which became valid on June 14, 1998, designates the Antarctic as a "natural reserve dedicated to peace and science." It bans drilling or mining for fifty years, and six "annexes" cover environmental impact, conservation of flora and fauna, waste disposal, marine pollution, and area protection.[48]

And now, as the Arctic sea ice is rapidly melting, corporations and nations are scrambling to obtain rights and permission to search for oil and other minerals. The result could be utter despoliation of this precious part of the globe. It is also more than ironic that global warming is inducing the Arctic ice melt while at the same time humankind wants to suck more oil from the Earth, which will exacerbate global warming—a vicious human-made circle!

Wildlife Smugglers

There exists a very lucrative international trade in wild animals and their component parts. As a member of the South Australian National Parks and Wildlife Council in the 1970s, I learned how our beautiful indigenous cockatoos were being drugged and smuggled in socks packed in suitcases aboard international flights. Australian snakes, birds, and marsupials are also part of this international commerce.

On a trip to Crete in 1987, I was walking along a side street when I heard a familiar scream—it was a sulfur-crested cockatoo looking very bedraggled and dirty in a cage outside someone's shop. Annoyed to see this smuggled bird, I grew even more incensed when I realized that it should, by rights, be flying free with a flock of hundreds of others in the Australian bush. How much had its captors paid for it? In all likelihood, $2,500 on the black market.

The international trade in wildlife and its products is now worth $4 billion to $5 billion a year (excluding fish and timber). Bangkok is used as an international transit station by dealers who "launder" illegally obtained wildlife, poached in Indonesia, Laos, Vietnam, and Cambodia. This international Mafia is threatening the extinction of the great panda, crocodiles, alligators, snakes, cacti, and orchids. Some years ago, a traveler from Mali was intercepted in Paris, and his luggage was found to contain fifty pythons, twenty tortoises, twenty lizards, and several vipers. In March 2004, Australian postal authorities found a package on its way to Japan containing twenty-four Oblong turtles and one shingleback skink lizard packed tightly in socks. Thirteen of the rare turtles were already dead. They would have fetched $1,200 to $1,400 each; and the skink, about $4,000.[49] A Japanese tourist arriving in Bangkok had eleven rare monkeys jammed into a carry-on bag; five of them had suffocated. Monkeys' teeth are often extracted with pliers or cut with clippers to make their bites harmless, and leopards' fur is dyed black to make it look like a house cat's. Americans have created a huge demand in rare parrots and dangerous snakes, and the Japanese like monkeys.

Chemical Destruction of Wildlife

Some of the most important, yet seemingly insignificant, species in the world are threatened by toxic chemical sprays used on crops. These creatures are the worms, fungi, insects, and bacteria that maintain a healthy soil base and root system for plants. They form the base of the pyramid of the food chain. Bees and other insects that pollinate crops and disperse seeds are also undervalued, but they are vital to our survival. Native wild bees and insects are worth about $60 billion to the US economy annually, but the bees are threatened by the little-understood Colony Collapse Disorder, which has accounted for the loss of one-third of the domestic bee colonies, putting $15 billion worth of crops at risk. The bees are simply not returning to their hives.[50]

The postulated causes for this disorder are new systemic pesticides that are relatively safe for humans but that intentionally disrupt insect neurology, causing memory loss and navigation failure. A family of highly toxic chemicals named neonicotinoids are mass-produced by Bayer and applied to seeds, which travel organically through the plant and leave residues on the pollen.[51] Other possible causes are habitat loss, climate change, electromagnetic fields, exotic pests, or the stress of being trucked large distances to work for the pollination industry,[52] and the rise of vast tracts of monoculture crops that create floral deserts when not in bloom.[53] Bats, which represent one-quarter of all mammalian species,[54] also scatter seeds. Many people probably do not realize that 90 percent of the most valuable US crops, worth a total of $4 billion, are fertilized by insects, and the catch-22 is obvious. As described already, pesticides and herbicides used to protect the crops from predatory insects and weeds kill the very organisms upon which the crops depend, including, of course, the bees.

Wild birds, bats, and parasitic insects have another function. They eat insect pests and therefore act as natural insecticides. So nature is clearly best left to itself. It has all the built-in mechanisms and feedback loops to ensure its ongoing health and survival. To put it crudely, we humans just screw things up.

The Sea

World market forces have helped destroy the ecology of the sea. New efficient techniques and excessive fishing have so depleted some fish species that they may never recover. For example, in 1992 the cod fisheries off the coast of Newfoundland collapsed from over-harvesting and habitat destruction, and they have never recovered. Currently, the large open-sea fisheries of cod, salmon, herring, and eel in the Baltic Sea are in crisis, as are many more fishery areas in Australia and the rest of the world.[55] Market forces are also responsible for pollution of the oceans by toxic sewage and poisons, plastic disposal, radioactive pollution, acid rain, and oil spills. Much of oceanic life is still a mystery to us because we cannot really explore the deepest layers of the oceans. Strange and wonderful life-forms have, however, been dredged from great depths, but even these species are not immune to the danger of sunken nuclear submarines and land-based poisons that we tip into the sea.

Pollution of the sea by plastic kills large numbers of marine animals. The fish eat pieces of indigestible Styrofoam and plastic, which cause intestinal obstruction or blockage of the bowel, and they die a slow death from starvation. Seabirds and their young also eat plastic, because it resembles fish, and they suffer similar deaths. Birds and fish often are caught up and trapped in the conjoined plastic rings that hold together a six-pack of beer or soda cans. These rings strangle birds and fish. I once found a dead albatross with one of these disposable obscenities wrapped around its neck, on the Ninety Mile Beach in South Australia. What a symbol of the industrial consumerist age! The National Academy of Sciences estimated that up to two million birds, ten thousand sea mammals, and countless fish die each year in US waters because of internal damage caused by plastics.[56]

One area in the northern Pacific Ocean—called the Central Pacific Gyre—is known as the "trash vortex" because winds and sea currents have brought together a floating plastic mass the size of Texas. More than one million sea creatures are killed each year because of this human-made plastic mess. All the plastic in the ocean—bottles, CDs, radios, TVs, furniture, car parts, toys, bags—breaks down eventually into smaller and smaller pieces, but it never disappears. One pound of plastic will eventually be converted into a hundred thousand tiny pieces, which are consumed by jellyfish, which in turn are eaten by larger and larger fish, allowing the plastic to work its way up the food chain. In the trash vortex, for every pound of plankton we find 6 pounds (2.7 kilograms) of plastic. The United States also produces a hundred billion pounds of small plastic pellets each year—raw material for toys and other goods, but a common pollutant in the seas. These plastic pellets tend to readily absorb up to a million times more DDT (pesticide) and PCBs (carcinogens) than are found as background water concentrations, and then the pellets are eaten by fish. Seventy percent of our planet is water, but less than 1 percent of our oceans' habitats are protected.[57]

Another very serious form of pollution is induced by eutrophication, when large quantities of fertilizers rich in phosphate and nitrates enter the rivers, seas, and oceans after heavy rainfall. These chemicals fertilize and stimulate the growth of massive blooms of algae, which then die, sinking to the ocean bottom, where they are consumed by bacteria. These bacteria suck up large amounts of oxygen from the water, creating vast oxygen-depleted areas. This phenomenon kills fish, clams, worms, seaweed, and a host of other creatures. Ocean areas like these, known as "dead zones," are rapidly increasing in number throughout the world, as the already widespread use of fertilizers increases. At last count, more than four hundred dead zones exist in global coastal waters, including a huge one in the Gulf of Mexico.[58]

Fishing ships have become fishing factories. Instead of catching tuna with rods, fishermen now use huge, sophisticated fish-finding sonar, spotter airplanes and helicopters, and huge seining nets to scoop up tuna and all other associated fish from large volumes of the sea. Over 7 million metric tons of seabirds, turtles, nontarget fish, and marine mammals are caught and thrown away each year.[59] Hundreds of thousands of dolphins are also caught in these nets and die unnecessarily. Over 70 percent of dolphins and porpoises are threatened as "bycatch."[60] International quotas must be placed on certain fish species in order to protect them. Tuna, for example, should be regarded as a luxury item and harvested accordingly, in relatively small numbers.

Large, international fishing boats owned by China, South Korea, and Europe fly flags of convenience from other nations, staying at sea often for years, fishing, refueling, changing crews, and off-loading their catches to refrigerated boats at sea. Others trawl the bottom of the seabed with vast fishing nets, damaging coral and unsettling eggs and fish breeding grounds, thus devastating the local industry. Fish is now the most traded animal commodity on the market, with a turnover of 100 million tons per year and Europe as the largest market. Fifty percent of fish stocks are exploited, and 25 percent are overexploited. A thriving illegal trade exists. Fifty percent of the fish sold in Europe is laundered like contraband, caught and shipped illegally beyond the limits of government quotas or treaties. Of the 100 million tons of fish caught yearly, about 30 million tons are discarded.[61]

The Japanese and Taiwanese, whose diets consist almost solely of fish and rice, are profligate fishermen. At present, they deploy drift nets, which are colorless, finely meshed nylon and about 35 miles (60 kilometers) long, that hang about 50 feet (15 meters) below the surface of the sea. The nets catch anything swimming in either direction along this barrier. They also trap seabirds that dive into the water to catch fish, and they ensnare whales, dolphins, sharks, turtles, and seals, as well as the target fish, salmon and tuna. The North Pacific drift-net fishing operation kills on average eight hundred thousand seabirds a year alone (not including those that fall out of nets before being brought on board).[62]

In effect, these nets are strip-mining the seas. In 1989, they netted 20,000–40,000 tons of tuna in the South Pacific, instead of the usual 10,000–15,000 tons. As a result, albacore and bluefin tuna stocks of the South Pacific are threatened, and it is thought that many species will never recover.[63] Roughly three hundred thousand cetaceans (whales, dolphins, and porpoises) die each year because they are caught in nets. About a hundred thousand albatrosses perish yearly as they are caught in longline nets. (Researchers have had to make wild estimates on this data because the available information is so unclear.)[64]

Many nations of the world have registered outrage at these fishing practices. As a result of international pressure, Japan will now stop fishing with drift nets a year earlier than expected, but Taiwan will continue to use drift nets despite the protests.[65] Japan, it must be said, hopes to resume the use of drift nets in the near future, if it can negotiate certain points with South Pacific countries. So each victory becomes, in effect, only a holding action, until new efforts are made. Greenpeace estimates that sixty-four hundred dolphins and thousands of fish of no commercial value were killed during a three-month drift-net fishing season in the Tasman Sea. Because of international pressure, though, in 1992 Australia, New Zealand, and other Pacific nations signed a convention to ban drift-net fishing in their oceans.[66]

Symptomatic of humankind's damage to the oceans is an alarming increase of invasions by jellyfish, which are stinging swimmers and tourists and clogging up fish nets from Spain to New York, Australia, Japan, and Hawaii. The explosion in jellyfish populations reflects a combination of overfishing their natural predators, such as tuna, sharks, and swordfish; rising sea temperatures; and depleted oxygen levels in coastal shallows.[67]

Whales

The Japanese are still whaling, despite an international law that bans the killing of whales. In the five years between 1985 (when the international ban was declared) and 1990, the Japanese killed 13,650 whales.[68] The blue, sperm, and right whales have been hunted to commercial extinction, and the fin and minke whales are threatened. Their carcasses are cut up and sold for meat for sushi, sashimi, and other delicacies. The Japanese government has used propaganda to tell its people that whaling is an essential part of Japanese culture and diet. Those who are opposed are labeled anti-Japanese.[69] However, whale meat in fact provides less than 1 percent of the protein in Japan. Therefore, Greenpeace Japan launched a campaign proclaiming that those who oppose whaling are not anti-Japanese. Japan also kills over 650 whales per year for "scientific whaling" purposes—able to do so because of a loophole in the International Whaling Commission's moratorium on commercial whaling.[70]

Overall, more than twelve hundred whales have been hunted each year since 1986, when the whaling ban took effect. Iceland and Norway also continue to catch and kill whales. The Russians stopped whaling in 1987, but in 1990 they announced that they would take thirty fin whales and seventy minke whales a year from the Sea of Okhotsk—to broaden their "knowledge and understanding of the marine ecosystem." That means they kill whales to learn that the whales are becoming extinct. In addition, in 1996 the Russians announced that they would allow the Chukotka Inuit to hunt bowhead whales for their survival.[71]

Meanwhile, the US Navy uses low-frequency sonar waves in tests to detect "enemy" ships and submarines at great distances. But these sonar waves severely disrupt the navigational systems of whales, dolphins, and other acoustically sensitive creatures, inducing hemorrhaging in their inner ears, and often causing spontaneous beaching and death. Finally in August 2008, under judicial order, the navy was banned by a California judge to use the sonar off the coast of California. The navy agreed to protect segments of the oceans that are critically supportive of life, while allowing the ships to use their sonar in large areas of the Northwest Pacific and around the islands of Hawaii. This compromise is obviously a smoke screen, because the navy has no way of knowing where the whales and dolphins swim.[72]

Dolphins

Dolphins, which have very large brains, are highly intelligent and capable of communicating with human beings. They can count, repeat words in a primitive phonetic form, and communicate emotions when they have befriended people. In addition to being slaughtered as a side effect of the large-scale fishing industry, dolphins are dying spontaneously of mysterious causes, which may well be related to ocean pollution. In an incident that occurred some years ago, scientists reported that an examination of 260 dead dolphins that had washed up on the beaches in Spain in August 1990 revealed that they were suffering from a viral infection similar to distemper in dogs. But they also had liver lesions caused by toxic substances in their blood that apparently had entered their bodies from the contaminated water. Scientists believed that poisons in the sea were inhibiting the dolphins' immune system, thus making them more susceptible to viral infections. This mechanism is very similar to the pathophysiology of AIDS, in which the immune system is depleted by the AIDS virus and patients die of massive bacterial or viral infections and cancer.[73]

Tourists are encouraged to feed dolphins with fish on certain organized cruises in the United States, but this practice has been discouraged by the Center for Marine Conservation because the fish themselves may be infected with diseases that are potentially dangerous to the dolphins.[74] Bottle-nosed dolphins and sea lions have been used for sinister purposes. The Reagan administration spent $30 million in a clandestine program to train them to guard the Trident submarine base at Bangor, Washington. Although navy officials said that the dolphins and sea lions are somewhat unreliable during training and occasionally go "absent without leave or refuse to obey orders," they admitted that what the animals lack in discipline, they make up for in sonar and speed: "their sonar system is better than any radar and they can pick up objects with incredible accuracy."[75]

Although naval officials have claimed to be phasing out the use of dolphins and sea lions since the late 1990s, there are plans for their use until 2012. And the navy still plans to send thirty dolphins and sea lions to Naval Base Kitsap-Bangor in Washington State, despite protests that stress on the dolphins, which are not native to this area, may result from the 10-degree colder waters, not to mention the possibility that they will transmit diseases to the native orca.[76] The Naval Ocean Systems Center, in San Diego, is a sort of "boot camp" for a hundred bottle-nosed dolphins and twenty-five California sea lions that are trained to find mines and stranded swimmers.[77] And although funding for this program was cut, the training continues.[78]

Sea Turtles

Sea turtles are under threat in the South Pacific. Fiji exports about 4,500 pounds (2,025 kilograms) of turtle shells to Japan each year. This figure represents a total of two thousand hawksbill turtles. The trouble is that only the large, mature turtles are captured and killed, and since these mature creatures are the breeders, the future of the whole population is in jeopardy. Six of the seven known sea turtle species are in serious danger of becoming extinct, according to the IUCN.[79]

Turtles are fascinating creatures. William (my son) and I stayed on Heron Island, on the Great Barrier Reef in Queensland, several years ago. As we lay luxuriating on the sand, huge female turtles clambered up the beach from the sea, laboriously dug a hole in the warm sand with their flippers, and proceeded to lay several hundred eggs. When the eggs hatched at night, the tiny baby turtles dug their way through the sand and headed toward the nearest light, mistaking it for the moon reflected in the sea. In this case, the light was from streetlights, and the baby turtles became lost. In the morning, the beach was still alive with turtles, but seagulls snapped them up. Some eventually got to the water, where big fish waited to devour them. Thus, out of thousands of hatchlings, only a few survive. When the surviving few disappear into the ocean, no one knows where they go. They reappear years later as giant adults.

Species Diversity

Before my trip to the Amazon in 1989, I visited the Galápagos Islands, which are cared for and protected by the Ecuadoran government. This group of islands is situated on the equator, several hundred miles west of Ecuador. The islands intersect about five ocean currents, which not only influence the climate but also have played a crucial role in the evolution of life-forms on the islands. Although I expected the weather to be very tropical and hot, it was in fact tempered by the cool ocean current heading north along the coast of South America and the current heading south from the west coast of Canada and the United States. These currents had, over time, carried an interesting variety of wildlife from various parts of the world to the Galápagos.

On one of his voyages, Charles Darwin visited the islands, where he developed his ideas on evolution after noticing that a particular finch on one island had features subtly different from those of the same species on another island. These landmasses are just far enough apart that the birds could not fly from one to the other; hence, they evolved different characteristics in isolation from one another.

On our trip, we traveled nocturnally by boat from island to island, and each morning we awoke to a different exotic landscape. The islands are volcanic in origin, and one of them is indeed still an active volcano. Some of the islands consist of stark, barren basalt, like the surface of the moon, and most have no water supply. Plants are primitive and sparse, and tall cacti have evolved to escape the predatory giant land turtles. Because of the inhospitable landscape, humans have not inhabited many of the islands and hence have had little impact on the wildlife. Giant land turtles are unique to the place, having migrated over eons from the sea to the higher ground, where they learned to graze upon grass. In the nineteenth century, pirates used to fill their holds with live turtles, which could survive for a year without food, and they would then be assured of a fresh, wholesome supply of meat. Now all the wildlife of the islands is protected.

I have never seen another region so teeming with life. Literally millions of birds, nests, babies, and eggs covered some islands. We walked along narrow paths snaking between nests sometimes built right on the path. The birds just stood and watched as we walked. Red-footed boobies mated while blue-footed boobies stood in their nests, looking as if they had blue rubber feet. I walked up the side of a cliff and came face-to-face with an owl that stared at me without expression. We came across a pair of albatrosses in the midst of a ritualistic mating dance. These birds are monogamous, mating with one partner for life, but they court for about a year. The dance was beautiful to behold. They clacked their long beaks together and then waved their heads around in a lovely, sinuous movement. The event continued for hours. The birds are so big that they cannot fly without taking off down a self-made runway and launching themselves off the edge of a cliff.

Thousands of sea lions lazed like giant slugs over the sands. Big males guarded harems of a hundred females, and live babies, dead babies, and placentas were strewn across the sand. William lay on the beach to take a nap and awoke to look up into the faces of four sea lions that were scrutinizing him. Herds of smelly sea iguanas sunned themselves on black basalt rocks, and huge land turtles trundled through the grass, retracting their heads and making hissing noises if we got too near. These creatures were the only ones out of hundreds of varieties that gave any impression that they even noticed us.

Twenty years ago, the Florida Everglades also presented an ecological paradise that teemed and undulated with life, but real-estate development and human-made canals have whittled away the natural ecosphere, and urban runoff and pollution are destroying the rest.

Once upon a time, the whole North American continent abounded with millions of animals and birds, as did Europe, the Middle East, Australia, and all other regions of the world. Not only are we losing species, but we are killing millions of genetically unique individuals within single species. As someone recently said, cutting down a forest is equivalent to shooting the animals and birds that live within.

I refuse to contemplate a world devoid of diverse life-forms. Is our development so important and sacrosanct that we must destroy all other species in our drive toward domination of the planet? Such behavior is anthropocentric. Let us instead develop a sense of humility and a deep love for our fellow creatures, recognizing that their value is equal to our own.

7 Overpopulation

During the nineteenth century, some well-meaning people introduced several English rabbits into Australia. Rabbits are normally considered nice, gentle creatures. But these domestic rabbits escaped and, experiencing no natural predators and liking the environmental conditions, began to reproduce with a fury. By the 1950s, the land in many parts of Australia was covered with rabbits. They ate the grass that was grown to feed sheep and cattle, causing erosion. They ringbarked trees, and they consumed small trees and shrubs. We were at a loss to know what to do with this affliction, until scientists at the Australian Commonwealth Scientific and Industrial Research Organisation (CSIRO) discovered a flulike virus called myxomatosis, which is highly contagious and lethal within the rabbit population. This virus was released into the environment, and the plague of rabbits disappeared within a couple of years. Luckily, the virus remained contained and had no other side effects. We still see some wild rabbits in the bush, but they are now relatively well controlled.

Having outlined in the preceding chapter the possible plight of most other species on Earth, I will now concentrate on the human species. While others diminish in number, we proliferate by the millions each day, like the rabbits. We, like them, have no natural predators. Having tamed the natural environment, we now rule the roost. Furthermore, the Bible says we were given dominion over the Earth, and many believe it. To reiterate, in 1800 we numbered 1 billion; in 1950, 2 billion; in 1990, 5.2 billion; in 2000, 6.1 billion; and in 2008, 6.72 billion.[1] By the second half of this century, we could number over 9.1 billion.[2] Few of us really understand the havoc we are wreaking on perhaps the only life system in the universe, and few of us seem to care. As long as we have access to our creature comforts and can buy and sell what we need, why think about the future? It's too scary. Yet beneath the veneer of comfort, we know what is really happening.

I have always lived in the Western world—fifty-two years in Australia and eighteen years in the United States. I had never visited India, because I was too frightened to come face-to-face with the poverty and suffering and feeling absolutely impotent, knowing that there was nothing I could do to help these people. But in 1988, I was invited by a medical and scientific organization to lecture in thirteen Indian cities on the medical effects of nuclear power, and I decided to accept.

Although I had read about the developing world for years, no amount of intellectual understanding prepared me for the actual experience. The sights, sounds, smells, disease, and poverty engulfed me. Children dragging their torsos about on skateboards, using their arms to propel polio-paralyzed legs; lepers without fingers and noses, begging on every corner; dark-eyed children with blond hair made light from malnutrition; bandy-legged children with rickety, bent bones secondary to vitamin D deficiency; and severe cases of kwashiorkor secondary to protein malnutrition. The streets were like a pathology museum—never had I seen so much preventable disease. Drinking water came from streams polluted with animal and human sewage. Hygiene in our sense was nonexistent. Yet, mixed with the odors of disease and decay was the smell of pungent spicy Indian food, and the people were vibrant and independent. Sitting on a street corner, one could see the most amazing collection of cultures, costumes, religions, caste systems, and peoples gathered together, living and working in relative harmony.

As I traveled the country, I was distraught to discover that the money of the very rich people in India would suffice to house, feed, clothe, and heal most of the people in their country. These people, born and bred to wealth, diamonds and gold adorning their bodies, walk among the poor and dying with unseeing eyes, much as rich New Yorkers step over the bodies of alcoholics and schizophrenics lying and dying in the gutters of New York City (although this situation has improved recently because many of them are now in homeless shelters).

In Mumbai, a large proportion of the population dwells in tents and lean-tos constructed from scraps of garbage, perched on top of the massive rubbish dumps that circle the city. They make their living by scavenging any piece of rubbish that can be retrieved and sold. The air is pungent with the smell of rotting debris and, on a hot, humid day, almost suffocating. Yet from these hovels emerge the most elegant, dignified women, draped in colorful saris—thin and fat, old and young, all with a grace and beauty that defies description. Each family often owns only one sari, so when one woman goes out, the others stay home. Even in the face of great deprivation and poverty, human dignity manages to prevail.

It was with some difficulty that I tolerated this physical assault upon my Western senses, until at the end of two weeks I became mortified and angry. I ranted at the poverty and disease, at the seeming heartlessness of the rich, and at my sense of despair and impotence. I hated India and wanted to escape. But the feeling of desperation passed within twenty-four hours, and I became more comfortable. As my tolerance for the morbid conditions grew, I was able to look at the scene with a sense of detachment and a degree of psychic numbing, or denial, that gave me a modicum of comfort. I suppose that is how the human psyche adjusts to difficult situations.

In the south of India, where lush coconut palms and banana trees hang over beautiful canals and the air is warm and moist, a population explosion is in progress. The Muslims are attempting to outbreed the Hindus in a competitive, almost warlike fashion.

I heard about one Muslim man who boasted of having five wives and sixty children. I had never seen such a huge number of children, and it was obvious that if drought or famine struck this population, a large number would perish. On a small planet with finite resources, such reproductive behavior is altogether inappropriate. India currently has a total population of over one billion people and makes up almost one-sixth of the world's population.[3]

Organized birth control was attempted in the 1970s by Prime Minister Indira Gandhi and her government, but in an inappropriate way. Hundreds of thousands of men were sterilized by vasectomy, having succumbed to the incentives of free transistor radios and other goodies. But eventually the people rebelled against this coercion, and birth control is now again a big problem in India. Associating birth control with the dangers of sterilization (offered for both women and men), many people also opted not to take any contraceptives; thus the population grew while the resources continued to dwindle.

In 1994, India developed a New Population Plan (NPP) to encourage women's economic, educational, and social welfare in order to improve and promote greater education on contraception and to control population growth.[4] In 2010 and 2016, however, the NPP will fail to meet its target because of bureaucratic inefficiencies and lack of coordination.[5] China, by contrast, offers a good example to the rest of the world. Its population now exceeds 1.3 billion, but it has been kept in check over the last several decades by a law mandating that each family have only one child.[6] This law could be interpreted as an assault on peoples' civil liberties, but with the planet facing the results of exponential population growth, such laws may be necessary to ensure planetary survival. Unfortunately, Chinese society still prefers boy babies over girl babies, and female infanticide is not uncommon. With the world population now at 6.72 billion, 20 percent of the global population, or one in five people on the planet, is Chinese. India makes up 17 percent, or one in six. In China, tax incentives are offered and financial fines imposed if people do not comply. The results of this policy are quite fascinating. A walk down any Chinese street at dusk is like strolling into a scene from the movie *The Last Emperor.*

A three-year-old child struts along the footpath, followed by two adoring parents and four adoring grandparents.

In contrast, several male-dominated religions have dogmas related to human reproduction that are obsolete and play a huge role in the progressive overpopulation of the world. Procreation and sexual rituals have for centuries figured prominently in the Hindu tradition, and polygamy is integral to the Muslim faith. The Roman Catholic Church is run by men meant to be celibate. Yet Jesus did not lecture his followers on the subject of birth control, contraception, or abortion. Indeed, the prohibition on abortion became part of the Catholic teachings only some hundred years ago. In a world threatened with extinction because of overpopulation, the pope continues to exhort people to have more babies.

In Dublin, Ireland, I saw a scene that I think illustrates contemporary Catholic tradition: a young, unkempt woman looking harassed, staggering along in high heels, carrying a screaming toddler, followed by a husband pushing a pram containing a tiny infant, and, in their wake, six other young children of various sizes and ages. Abortion remains illegal in Ireland. However, contraception was progressively legalized in 1979 and 1984, and it is now fully legal. Moreover, Mary Robinson, who believes in birth control, abortion, and divorce, was elected president of Ireland in 1990, and she helped to create political change with regard to these important issues. Divorce is now legal.

It is clear that the reproductive policies of certain religions are out of date, if life on Earth is to continue. I am not suggesting that the personal quest for spirituality and enlightenment, as exemplified by the teachings of Jesus, Buddha, Allah, or the Hindu gods, is obsolete. But it is beyond time for women to assume leading roles in the Earth's representative religions and to establish new and more appropriate spiritual organizations. I do not make this statement lightly, but with a sense of gravity and urgency.

Fifty-two percent of the human population is composed of women.[7] According to the United Nations, women do 66 percent of the world's work, including much of the work of producing food, and for this labor they receive 5 percent of the global income and own only 1 percent of the property.[8] And they have very little power. Because women give birth to all the babies, they bear a major responsibility for the population explosion and are thus in an extraordinary position to stop the exponential increase of *Homo sapiens*. (The mathematical concept of exponential growth is not difficult to understand. For example, a water lily in the middle of a huge pond takes four weeks to reproduce. It then takes four weeks for another doubling, from two to four lily leaves. After ten years of steady growth, the pond is half full, but the next doubling of four-weeks duration will fill the pond.)

It is a well-known fact that in places where women are well educated and relatively affluent, the birthrate is low. Because they have access to contraception, they understand the reproductive cycle and can therefore control their own destinies. Poor women tend to have many babies, because they are never sure whether their children will survive the diseases and malnutrition of infancy. Furthermore, poor parents need adult children in their old age as a form of insurance—to grow food, to run the farm, and to care for them. About twenty-six thousand children under the age of five die each day from malnutrition and preventable disease like diarrhea, measles, tetanus, whooping cough, polio, and diphtheria (which accounts for 53 percent of the overall deaths).[9] Eighty percent of all disease in the developing world is caused by a lack of access to clean water. (In fact, community health is measured better by the number of water taps in a village than by the number of hospital beds.) Some 1.1 billion people do not have access to clean water, and 2.6 billion do not have adequate sanitation.[10] Malnutrition predisposes infants and children to contagious diarrhea and other infectious diseases—hence the very high death rates among children in developing countries, where 1.4 million die each year from lack of clean water.[11]

Contraceptives of all varieties must be made available to those of reproductive age. Every person has as much right to birth control as to food. According to the United Nations, over half of the hundred million married women in developing countries, excluding China, need contraceptives because they do not want more children. Over one-third of pregnancies are unintentional, and of these, two-thirds are the result of not having used contraception.[12]

Yet most of these women are poor and uneducated and have not been taught about the ovulation cycle, let alone how to use contraceptives safely, even if they had them.

Vasectomy is not popular among men, although a one-year study by the World Health Organization found that weekly injections of a synthetic male hormone "can maintain safe, stable, effective, and reversible contraception."[13] Some contraceptives are particularly useful for people in developing nations. For example, injectable Depo-Provera is a long-term hormonal suppressor of ovulation, as is a preparation called Norplant, which can be implanted under the skin and remain effective for five years. The debate about abortion in the United States strikes me as obsolete and misogynist. Thirty million species are endangered by our relentless procreation, yet we still argue about abortion as if we lived two centuries ago, when the human population stood at one billion.

So invidious is this absurd American debate that during the Reagan years, funding for overseas birth control aid was cut off to countries that permitted abortion—including, of course, almost the entire developing world. This policy was known as the "Mexico City Policy," or more commonly the Global Gag Rule, and it ended with the Clinton administration in 1993. The policy restricted foreign nongovernmental organizations (NGOs) that receive USAID (US Agency for International Development) family-planning funds from using their own, non-US funds to provide legal abortion services, lobby their own governments for abortion law reform, or even provide accurate medical counseling or referrals regarding abortion. Ominously, on the first day of George W. Bush's first term, the Global Gag Rule was reinstated—ironically on the twenty-eighth anniversary of *Roe v. Wade*.[14]

Bush also added bans on funding organizations that teach abortion education or any political activism related to limiting family growth. In 2002, Bush withdrew foreign aid worth $34 million that was to have been used by the United Nations Population Fund (UNFPA), the largest provider of reproductive health services in the developing world. UNFPA estimated in 2002 that this cut would result in "2 million unwanted pregnancies, 800,000 abortions, 4,700 instances of maternal mortality and the death of

77,000 children under the age of five."[15] A new, quick-and-easy contraceptive is a drug called RU-486. It is a hormone developed in France that, if taken within days or weeks after conception, induces a painless, spontaneous abortion. Years of testing have revealed virtually no serious side effects.[16] The Clinton administration granted RU-486 its legality. If I could, I would like to be a world salesperson for RU-486—what a difference this easy form of birth control could make to overpopulation!

Compare the minuscule sums that these programs require with global military spending. President Reagan spent more money than all past presidents combined, and most of it went to the military-industrial complex for high-tech weapons systems, many of which were used in the Persian Gulf War. In 2004, while US soldiers were lacking armor and running low on supplies in Iraq, the Department of Energy spent $6.5 billion on nuclear bombs and warheads. Adjusted for inflation, that's equal to what Reagan spent at the height of the tension between the United States and the Soviet Union. And the cold war is over![17]

World military and arms trade spending in 2005 was greater than $1.1 trillion, a 34 percent increase in a decade. The United States accounts for 48 percent of this amount.[18] According to the Worldwatch Institute, if the much smaller sum of $25 billion a year went to social improvements in the Third World to reinforce birth control policies, the world population would stabilize at eight billion by 2050. Improved socioeconomic status would lead to better prenatal care, lowering maternal and infant mortality. Third World maternal deaths, which number half a million a year, are related to malnutrition, prenatal anemia, and concurrent infectious disease.[19]

Almost half the abortions in the world are illegal (61 percent of the global population lives in countries that have legalized abortion, and these countries have fewer people),[20] and women who have to resort to illegal abortion often die dreadful deaths from gram-negative septicemia, or blood poisoning, resulting from dirty instruments. Deaths are rarely associated with legal abortions, but in countries where abortion is illegal, rich women who can afford sterile, professional procedures always get safe abortions, while poor women often die. Years ago when I worked in the casualty department of the Royal Adelaide Hospital, a woman was brought in on a stretcher, desperately ill from an illegal, septic abortion. She was accompanied by a weeping husband and six terrified children. She was going to die and would leave her family behind her—all because abortions were not legal at that time and because she could not cope with more children.

In my years in general practice, it was often the Catholic women who knocked on my door when their thirteen-year-old daughters were pregnant, or when they themselves had become pregnant by a man who was not their husband. It was clear that religious principles often went by the wayside when the practical issues of life and death presented themselves.

Further, urgent global educational efforts related to sexually transmitted disease are needed. AIDS is spreading like wildfire in Africa and elsewhere. The sexually spread wart virus is related to cervical cancer, vulval and vaginal herpes can cause herpes encephalitis in a newborn baby, some forms of syphilis are drug-resistant, and gonorrhea is a nasty disease often causing arthritis, pelvic inflammatory disease, and adult blindness.

It may seem somewhat contradictory for me to stress, in discussing overpopulation, that AIDS and other diseases must be prevented. Yet the cure for overpopulation is not epidemics of disease, or nuclear war, as some people suggest. It is redistribution of wealth, compassionate politics, and caring societies. Logically, if women and men are well fed, well educated, and financially secure, their children will not die in infancy and the birthrate will automatically drop because they will expect their children to survive.

Cuba is an example of a country with good medical care and an equitable distribution of wealth. Before the US-backed dictator Fulgencio Batista was overthrown by Fidel Castro, Cuba was controlled largely by US business interests, the Mafia, and wealthy American tourists. When I visited Cuba in 1979, I learned that before the revolution, the majority of the population had been severely repressed, illiteracy had been endemic, and more than 70 percent of the people had suffered from anemia, hookworm, intestinal parasites, and tuberculosis.[21] The introduction of communism certainly brought with it political repression and infringements of civil liberties, but the population at large was liberated. Several years after the revolution, almost all adults were literate, all children attended school, prenatal clinics had been established, and prenatal care was mandatory for all women. Infant and maternal mortality rates plummeted, ranking among the lowest in the world. Free medical care became available to all, and as the standard of living increased, the birthrate fell.

My visit to Cuba was a wonderful experience, but I felt distressed that ever since the revolution the United States had applied a trade embargo on sugar exports and on such imports as medical drugs, machinery, oil, and other necessary items. Despite these restrictions, Cuba demonstrated that a small, isolated, depressed country could raise the standard of living for its people.

We were free to wander through city streets and country roads, and we talked to many people. One old peasant woman broke into a huge grin when I asked her what she thought of the revolution and of Castro. She said with enormous pride, "My son's a doctor." He would never have received that educational opportunity while the country was under Batista's rule.

Bearing in mind the conditions of the developing world, we must be acutely aware that the problem of overpopulation will eventually also become critical in parts of the developed world that are still reproducing beyond a stable population base. For instance, the US population, which was expected to reach 268 million by 2000, had already grown to over 282 million in 2004. The US Census Bureau expects 49 percent growth in the next fifty years, while the populations of most European countries are expected to decline in those fifty years.[22] The United States is growing faster than eighteen other industrialized nations.

All of us must therefore accept responsibility for our reproductive habits. No longer can we say that it is God's gift that we are pregnant or that it is okay for us to have babies because we are wealthy and our population growth rate is lower than that of many Third World countries, while we exhort the developing world to slow down. Those attitudes are seen as racist and elitist. We must all take responsibility and realize that we are suffocating thirty million other precious species by our egocentric and anthropocentric attitudes. It is time for the global community of *Homo sapiens* to take the radical step of limiting families to one child. This policy could be formulated by the United Nations, so that the nations of the world could thereby assist and aid one another while taking responsibility for their own societies. Of course, birth control must be voluntary on the part of educated individuals, endorsed and supported by government subsidies and help, rather than mandated by law. Such behavior could become fashionable and an integral measure of responsible societies after appropriate national and international educational regimes.

Women are, in all ways, crucial to planetary survival. In developing countries, they do most of the agricultural labor, they understand the rotation and planting of specific crops, and they accept the responsibility to provide good food for their families if they can. In sub-Saharan Africa, for instance, women produce four-fifths of the food. Yet it is the men who own all the tools, production facilities, roads, trucks, and land and who make all the key political decisions. Because the contribution of women is almost totally unrecognized, they are officially said to make up 20 percent of the agricultural work force.[23] It is true, however, that in 2004 Dr. Wangari Muta Maathai, of Kenya, was the first African woman to receive the Nobel Prize, for "her contribution to sustainable development, democracy and peace."

In Nepal and India, it is the women who slave in the fields and who carry huge sacks of rice slung from burlap straps around their heads, while the men tend to sit around drinking tea and coffee, philosophizing about life. Similar scenes are frequently seen in TV shows that portray life in African countries.

Several stories told in *The Global Ecology Handbook* illustrate the hazards involved when women's work is ignored.[24] World Bank reforestation projects in Kenya and India failed when the education and advice were directed toward men, but succeeded when women were taught to care for the trees. In Gambia, women's control of development projects was initially prohibited, yet when women were given land rights and provided with seed, equipment, and centers for child care, rice production increased sixfold.

According to the 1987 United Nations report *Our Common Future*, women are the single most important sector of society when it comes to lowering population growth and caring for the Earth. We ignore women's contribution to world politics at our peril. It is time for men in religion, in politics, in corporations, and in global agriculture to stand aside and attest to the intelligence and wisdom of women. The birthrate will fall, wealth will be redistributed, compassionate societies will evolve, wars will almost certainly cease, and food will be efficiently produced and distributed. I write generically about women, of course, because specific examples, such as Margaret Thatcher, do not represent the true attributes of the large majority of sensible, wise women.

8 First World Greed and Third World Debt

I place economy amongst the first and most important republican virtues, and public debt as the greatest of the dangers to be feared. To preserve our independence, we must not let our rulers load us with perpetual debt.

—Thomas Jefferson, third US president, architect, and author (1773–1826)

When I think of malnutrition and of twenty-six thousand children around the world dying daily from starvation-related disease, I am brought brutally face-to-face with the affluent countries.[1] The gap between the 20 percent rich and the 80 percent poor is not decreasing but increasing. Ten percent of the world's richest people control 85 percent of the global wealth, and 30 percent of that wealth is found in the United States. The world's population is 6.72 billion, and about 1.1 billion live on less than $1 a day. According to the World Bank, the average annual per capita income in the developed countries is $17,000.[2]

Access to food is a preoccupation of 80 percent of the world's people, yet food in the United States is overabundant, and many Americans consider it a difficult substance to understand and sometimes even poisonous. It is true that fruits, vegetables, and processed goods are often contaminated with toxins, pesticides, artificial dyes, hormones, and antibiotics. On the other hand, average life expectancy continues to rise in the United States—it is now 77.6 years of age—because of a combination of preventive medicine, medical care, and adequate food; while in the Third

World it progressively declines, and in many countries it is in the mid-thirties. Because of obesity, however, the extraordinary life expectancy in the United States is expected to fall in the coming years—by about two to five years within the next fifty years.[3] The people of the First World—Americans especially—are obsessed with vitamins, cholesterol, sugar, and the like, while the majority battle obesity or, to put it simply, overnutrition. In 2004, it was estimated that 820 million Third World people, by contrast, did not consume enough calories for a normal active working life.[4]

I am repulsed when I go out to dinner and read on the menu, "All you can eat for $22" and am served a steak weighing one pound and a huge potato replete with sour cream, a salad with a choice of ten salad dressings, and bread soaked in garlic butter. I know that more than half the food will be thrown away, because my stomach's capacity is too small and I do not need all that food, while twenty-six thousand children die daily of starvation. Fourteen percent of the food used by US homes and restaurants is thrown away, and this food is worth $43 billion per year.[5]

In fact, people in wealthy countries need to eat less, to eat items lower on the food chain, and to share their riches. The grossly excessive amounts of food produced in the United States should be exported to Third World countries that really need it. (Twenty-five percent of children in the United States are obese, and of that 25 percent, 80 percent are likely to become obese adults.)[6] But conditions must be imposed on exporting activities so that Third World farmers are not jeopardized by the dumping of cheap imported food on their markets.

Apart from the poor people who live on the fringes of US society, most Americans probably have ample or excessive amounts of vitamins, proteins, carbohydrates, and fats in their diet. We must be aware, though, that in the wealthiest country on Earth in 2002, 34.6 million people were malnourished because of inadequate or inappropriate food intake.[7] By 2007, approximately thirteen million children, or 18 percent of US children, lived below the poverty line, which was classified at that time as $20,000 a year for a family of four; and between 2000 and 2005, the number of poor children increased by 11 percent.[8] But from a global perspective, it is immoral that a small minority of people in the world are overnourished, while most are undernourished. We should also understand that severe malnutrition in childhood induces mental retardation, because the developing brain needs to be well fed. This means that millions of people are unable to better their condition, because their mental capacity has been damaged by poor nutrition early in life. And there is absolutely no reason why the wealthiest country on Earth cannot feed all its people. Civil society must pressure the government to make policy changes to correct this situation, which eats away at the moral fabric of American society.

Studies reveal that Americans eat twice as much protein a day as they need, and it is clear that many common diseases in the US population are related to the problem of overnutrition.[9] For instance, excessive or daily consumption of red meat over a long period of time increases the incidence of colon cancer in women and probably in men as well;[10] it also induces high cholesterol with its associated hypertension, atherosclerosis, heart attacks, and strokes, as well as breast cancer.[11] Moreover, the obesity caused by overeating can lead to hypertension, heart disease, strokes, diabetes, osteoarthritis, and joint conditions. Hypervitaminosis can result from excessive doses of vitamins A and D, and many Americans take large quantities of extra vitamins, when, in fact, they already obtain their daily requirement of vitamins.

What can Americans do to make sure that they are nourished better and that their eating habits can help benefit the global population?

- *Eat foods lower on the food chain.* A high intake of fruits and vegetables ensures both a high-fiber diet, which decreases the incidence of colon cancer, and a low-cholesterol diet.
- *Eat less meat.* Most red meat and some chickens are contaminated with antibiotics, the same drugs that doctors prescribe to kill bacteria in infected human patients. Widespread, indiscriminate antibiotic consumption by the general population leads to drug-resistant bacteria, nullifying important therapeutic approaches to serious human bacterial infections. Ste-

roidal hormones, including bovine growth hormone, are also fed to cattle, in order to increase the weight and muscle content of the meat, thus maximizing the farmer's profit. Meat also has a high saturated-fat content, which often leads to arteriosclerosis, heart attacks, and strokes.

- *Eat more grain.* In the early twentieth century, Americans obtained 40 percent of their protein from grains and cereals, but now only 20 percent comes from such products. Hundreds of years ago, when our forebears lived as tribal nomads, the diet consisted mainly of fruits, vegetables, and grains. But with the advent of more "sophisticated" lifestyles, the consumption of animals increased. In biological terms, our bodies did not evolve to depend on large quantities of animal protein—it is not good for us! Furthermore, just as the world environment cannot sustain 6.72 billion people, it can hardly sustain the present global population of 1.3 billion cows, 1.7 billion sheep and goats, 85 billion pigs, and over 10 billion fowl.[12] It takes 13 pounds (about 6 kilograms) of grain and soybeans[13] and 435 gallons (about 1,650 liters) of water to produce 1 pound (about half a kilogram) of beef, 6 pounds (2.7 kilograms) for 1 pound of pork, and 3–4 pounds (1.4–1.8 kilograms) for 1 pound of poultry or eggs. Animals now eat five times more grain than do the American people, enough to feed the entire US population six times over.[14] The production of each acre (a little less than a half hectare) of food uses energy equivalent to 500 gallons (about 1,900 liters) of oil. Of the total US energy consumption, 14 percent goes toward food transport, 16 percent to food processing, and 7 percent to packaging food. In 1990, the 145 million tons of grain fed to animals yielded only 21 million tons of meat, chicken, and eggs.[15] By 1997, 40 million tons of plant protein was used to produce 7 million tons of animal protein. For every 2.2 pounds (1 kilogram) of animal protein, the livestock was fed 13.2 pounds (6 kilograms) of plant protein.[16] What an extraordinary waste of soil, water, fertilizer, pesticides, and energy—and all for what return but increased heart attacks, obesity, and strokes?

Although China is now the number one consumer of fertilizer, using more than 40 million tons in 2004, the average American uses an estimated 530,000 gallons (about 2 million liters) of water per year, 80 percent of which is allocated to food production.[17] On a daily basis, the United States consumes the equivalent of 2,200 pounds (about 1,000 kilograms) of food per person.[18]

The Feinstein World Hunger Program at Brown University estimates that if the world population is fed primarily with grains and vegetables, there is, at present, enough food to ensure the USDA recommended daily per capita intake of 2,200 calories for the global population.[19] But while billions starve, about two-thirds of the grain grown in the world and half the fish caught are fed to animals in rich countries. If the world's population reaches eleven billion, its annual food production must increase two and a half times merely to maintain the current situation and low per capita output.[20] We have work to do!

About 60 percent of the world's best agricultural land is situated in only twenty-nine countries, and over 40 percent of the world's agricultural land has been badly degraded.[21] A mere 11 percent of the land area of the planet is suitable for agriculture, yet soil erosion associated with deforestation, dam construction, and irrigation is depleting 75 billion tons of the world's topsoil each year.[22] Our global treasure is being lost in the sea and silting up rivers. In addition, about 2.5 million acres (1 million hectares) of rich farmland in the United States—an area more than double the size of Delaware—is destroyed by "development," covered with asphalt, freeways, houses, and shopping centers.

In a world where equitable redistribution of wealth is vital, let's now discuss the financial support that the wealthy 20 percent currently give to the poor 80 percent. In 2007, the United States spent over $625 billion on its military, while, in contrast, it contributed only $22.7 billion in foreign aid in 2006. Unfortunately, only $2 billion went for food, $3.2 billion for humanitarian aid, and $5.6 billion for sub-Saharan Africa.[23] Of the $22.7 billion in US government foreign aid in 2006, Israel received the largest amount—$2.5 billion, most of which it spent on its military—Egypt received the second-largest amount, followed by Iraq and

Afghanistan; and only four of the world's poorest nations—Ethiopia, Liberia, Uganda, and Tanzania—were among the top twenty listed aid recipients.[24] US citizens, however, tend to be more generous than their government, giving $34 billion in private overseas donations in 2006.[25] Note the radical difference between spending money on killing versus spending money to help needy people throughout the world. The most needy countries were and are continually shortchanged.[26]

Most aid serves as an instrument of foreign policy, not really as a charitable gift. For example, during the years 1991–2003, the United States organized through the UN to impose sanctions on Iraq, depriving it of necessary food imports and medical drugs, which helped to lead to the deaths of five thousand children from malnutrition and waterborne diseases each month. Similar sanctions involving food and medical drugs have been imposed on Cuba since its revolution in 1953, and these sanctions continue to apply today. These are just two instances in which the US government has withheld food for political purposes. Food is used to reward and manipulate poor countries rather than to feed hungry people.

Surprisingly, most US aid actually winds up subsidizing American corporations. Between the years 1995 and 1999, the United States gave between $6 billion and $15 billion in foreign aid, 80 percent of which went to US companies in foreign countries. The main beneficiaries were Israel, Egypt, Pakistan, India, and the Philippines.[27] So US foreign aid serves not only as a coercive instrument of foreign policy, but also to support private US contractors, universities, banks, consulting firms, lobbyists, and so forth. In fact, foreign aid is now recognized to be a lucrative business, and companies are scrambling to capitalize on it. Corporations also tend to borrow most of their investment funds for Third World projects from Third World banks.[28]

In 2006, when US foreign aid expenditure totaled over $22 billion, Americans spent $9.5 billion on movies and theaters,[29] $86.6 billion on tobacco (in 2003),[30] and $155 billion on alcohol,[31] plus $35 billion on weight loss items that don't usually work.[32] An expenditure of just seven cents per person would save the sight of

350,000 children who are blinded annually because of a vitamin A deficiency,[33] and a mere $3 each would immunize them against poliomyelitis, tetanus, whooping cough, diphtheria, and measles. One year's expenditure by the US cosmetics industry would provide 1.6 billion people with sanitation.[34] Providing water and sanitation to some of the poorest countries would cost an estimated $9 billion annually, less than Europe's annual expenditure on ice cream.[35]

In percentage of gross national product (GNP) given away as foreign aid, Sweden topped all countries, with 1.03 percent in 2006; and Norway, which dropped from first to third, spent 0.89 percent. The Netherlands spent 0.81 percent of its GDP. The United States ranks next to last out of eighteen countries, with 0.17 percent of its GNP and is almost always among the lowest from the industrialized nations of the world. Yet even the best of these figures is still very small.[36] How does the United States spend its money? In 2005, the United States ranked first in military spending, arms exports, military technology, naval fleets, global military bases, and number of nuclear bombs; while at the same time it was eighth in adult literacy rate, ranked fifty-first in per capita spending on secondary-level public education,[37] and had the highest newborn mortality rate in the developed world.[38] These data, though not strictly financial, indicate present priorities in American society.

The United States spent $626.1 billion in 2007 on weapons, military personnel, and military equipment, while the total global expenditure on these items is $1 trillion.[39] In 2006, the Department of Defense (or, more aptly, the Department of "Offense") received the largest increase in funding since the Reagan administration, and a 41 percent increase over the amount spent in 2001.[40] One-third of the more than $70 million that is spent on a single Trident submarine could eliminate malaria in the world by clearing mosquito-ridden waterways and providing prophylactic medicines to the target population, and one week of global military spending could put an end to all starvation.[41] The war in Iraq is expected to cost from $3 trillion to $5 trillion in toto! The good that that money could do to alleviate poverty and disease is obvious.[42]

To quote Thomas Jefferson, "The care of human life and happiness, and not their destruction, is the first and only legitimate object of good government." Or as Dwight D. Eisenhower put it, "Every gun that is made, every warship launched, every rocket fired, signifies in the final sense a theft from those who hunger and are not fed, those who are cold and not clothed."[43]

International aid is but a Band-Aid on the wounds of Third World suffering. The people there are not just malnourished and deprived because of overpopulation, inadequate distribution of money, lack of education, or bad land management. They are poor and starving because financial powers in the developed world exploit them to satisfy their own greed and continued affluence.

Two major dynamics occurring globally are having, and will continue to have, severe and lasting repercussions on the developing world. One is the huge foreign debt incurred by the developing world, which increased from $618 billion in 1980 to $3.15 trillion in 2006, according to figures published by the International Monetary Fund (IMF). The IMF predicted that the external debt of this group of countries, comprising 145 member states, would continue to grow throughout 2007, to more than $3.35 trillion.[44] The other dynamic is the dangerous set of negotiations—known as "free-trade agreements"—that took place in the organization called the General Agreement on Tariffs and Trade (GATT), which was replaced in 1994 by the World Trade Organization (WTO).

Now, I recognize that some of this information can appear boring and dry, but these economic realities will shape the destinies of most of the world's people and thirty million species. We are therefore obliged to study hard, think, and act.

Third World Debt

Third World debt is exacerbating global environmental degradation. Until 1973, the poor developing countries had made great strides in public health and preventive medicine, in education, and in crop production. But in the years 1973–74 world dynamics changed. The oil-producing, OPEC countries suddenly increased the price of oil fivefold, and oil became scarce. There were large and often angry lines outside gas stations in the United States and elsewhere. The oil-rich countries made huge profits and deposited their money, called "petrodollars," in the world's major banks, particularly in the United States.[45] Between 2002 and 2007, crude-oil prices increased almost fivefold again. The trading price of a barrel of oil in 2002 was under $20, and by November 2007 it had reached almost $100.[46] In August 2008, it was $143 a barrel, but as I write now in December 2008 it has fallen below $50 a barrel, largely because of the global recession. These figures reflect a decreasing supply of readily acquired oil and an increased global demand, combined with market speculation on future oil prices.

The rising oil prices in the 1970s injected large amounts of extra cash into banks, which were eager to lend it out in order to profit from this windfall. So the banks decided to offer low-interest loans to Third World countries, which could obviously benefit from extra cash. The banks called this lending policy "recycling," but it turned out to be, in effect, an international lending spree, in which they unloaded petrodollars on unsuspecting developing countries.

The World Bank, which represents many of the major banks in the United States and other wealthy countries, decided at a conference in the Philippines in 1976 to lend this impoverished country $4.5 billion. And by 2000, the World Bank had loaned the Philippines $10.3 billion, including $361 million from Japan. (The money is mainly directed toward the island of Mindanao, which has a large Muslim population, with the hope that the government can control the Muslim community.)[47] The World Bank openly stated that it would rather lend to "stable" countries (that is, countries ruled by dictators) than to democracies governed by the people. Hence, President Ferdinand Marcos of the Philippines was an excellent client in the mid-1970s. Today the IMF and the World Bank still regard the Philippines as a "favorite guinea pig." The Philippines is on the top-ten list of the world's most indebted countries, and it struggles to receive aid. In the mid-1990s, only 10 percent of the population increased their income, while 90 percent saw a loss of income; and the top 20 percent earned fourteen times more than the lowest 20 percent.[48]

Countries with similar dictatorships at that time also received huge petrodollar loans—for example, Chile, Brazil, Argentina, and Uruguay. Other borrowers with suitably compliant governments included Mexico, Tanzania, and many struggling African nations. The amount of money available was vast. Lending to Latin America increased from $35 billion in 1973 to $350 billion in 1983.[49] By 2005, Latin America owed $350 billion in loans, and in fiscal year 2007 the World Bank loaned Latin America and the Caribbean $4.6 billion. Argentina, Brazil, Colombia, and Peru received the largest amounts.[50]

Most of the loans were based on variable interest rates. During 1981–82, US interest rates doubled to 20 percent because of tight monetary policies caused by the growing US deficit, engendered by the Reagan administration's lavish spending on weapons. But each 1 percent increase in the interest rate meant $4 billion more that Latin America had to give US banks in interest payments. Furthermore, a global recession at that time decreased the demand for exports from developing countries, thus decreasing their income. Debt servicing rose from 15 percent of their export earnings in 1980, to 31 percent in 1986 in sub-Saharan Africa. Between 1973 and 1980, Third World debt increased by a factor of 4, to $650 billion, and by 2007 it had risen to at least $2.9 trillion.[51] This is an unbelievable burden for countries whose populations are barely surviving. The developing world must spend $13 on debt repayment for every $1 it receives in grants. About sixty of the world's poorest countries have paid $550 billion in principal and interest in the past three decades, yet a $523 billion debt burden exists.[52] By 2005, sub-Saharan Africa owed about $200 billion in loans—a huge 83 percent of its GDP.[53]

To make this situation more graphic, imagine that you accepted a variable-rate mortgage (VRM) of 8 percent in 2007 for $100,000 to buy a house. But instead of being "capped," as most VRMs are, at an arbitrary level—say, 15 percent—yours had no legal ceiling. When interest rates went up, therefore, so did the interest on your mortgage—from 8 to 20 percent! But at the same time, your income failed to increase, or even decreased. Instead of $800 a month in mortgage payments, you were forced to pay $2,000 per month. You would quickly become bankrupt. This was the plight of many Third World countries. Ironically, a similar situation has developed in the United States with the subprime crisis initiated by irresponsible lending on the part of banks and credit companies to people who could clearly not afford to pay the interest accrued on the loans. In the meantime, interest rates went up, property values went down, and many hundreds of thousands became bankrupt and lost their homes and their savings. Banks were also affected, many teetering on the brink of disaster. The US economy is now in recession, or even depression, because of this wave of totally irresponsible lending.[54]

Loans to other countries have been used to finance badly planned highway systems in big cities inhabited by the rich, as well as railroads, dams, and power plants. These projects had no relevance to the millions of people trying to sustain an existence with poor land, archaic equipment, minimal housing, and few, if any, health care facilities. And all profits accrued to the foreign corporations that carried out the construction.[55]

The World Bank has encouraged countries to destroy tropical rain forests to pay back their debt. In Brazil, it proposed that about 12.5 million acres (5 million hectares) of the Amazon be brought under control and management. Early in 2008, Brazil's second largest exporter of beef won approval from the International Finance Corporation (IFC), the private equity lender of the World Bank, for a $90 million loan. The IFC said that the project is "an opportunity to partner and engage a leading private sector company in tackling the most serious environmental and social issues facing the Brazilian Amazon."[56] So far, more than 60 percent of the clearing carried out in the Amazon forest has been for cattle pasture, much of it ultimately for hamburgers for US consumers and soybean production to feed animals in Europe. The World Bank has also encouraged deforestation in Ecuador (about 2.5 million acres, or 1 million hectares); in India (about 74 million acres, or 30 million hectares); and in Papua New Guinea. In addition, the World Bank funded the Democratic Republic of the Congo's destruction of much of its rain forest and implementation of "zones" for locals to cultivate.[57] It claimed that industrial forestry would contribute strongly to the country's economy after years of war. Congo's rain forests are the second largest in the world after the Amazon's rain forests, and they lock up 8 percent of the planet's carbon and are rich in biodiversity. Forty million people depend on the forests for medicines, shelter, timber, and food, and between 250,000 and 600,000 pygmies are threatened by this deforestation.[58] The World Bank also approved a loan of $156 million for a dam on the Serang River, in central Java, Indonesia, which would entail the resettlement of twenty thousand people and the flooding of huge areas of forest. Similar projects were orchestrated in Zaire and in Gujarat, India.[59]

After the Mexican crisis in 1982, the IMF decided to impose severe austerity measures that forced desperately poor Third World countries to use more and more of their land to grow cash or luxury crops for export, such as bananas, coffee, cocoa, pineapples, and flowers, in order to help pay off the debt. Forests and virgin land are being destroyed and cultivated by corporations and large landholders, producing ecological damage and leaving very little land on which the indigenous population can grow food. The people then become dependent on cheap imported food from the United States (grown with subsidies from the US government). At the time of the decolonization of Africa in the 1960s, Africa was a net exporter of food. Now it imports 25 percent of its food.[60] The net effect of these policies has been to produce nations that are hostage to the American and European agricultural systems and that lose their independence. Instead, foreign aid should be used primarily to help Third World countries establish agricultural and economic autonomy.[61]

These severe austerity measures by the IMF and World Bank were initiated in 1983, when every country that had applied to these institutions for loans was told to revamp its economy from top to bottom with a list of economic policies that was considered a bare minimum for economic health. The list included such overtly contentious issues as these:

- All state enterprises should be privatized.
- Barriers that impede the entry of foreign firms should be abolished.

These dictates were derived from the philosophy of a very conservative and dangerous economist, Milton Friedman, whose basic nostrum for healthy economies consisted of privatization of state enterprises, deregulation (or no state or government control over corporations), free trade, and drastic cuts to government spending. Friedman was followed religiously by American economic gurus, but his theories and philosophies have done extraordinary damage to societies both in the United States and in the Third World. Countries such as Chile, Argentina, Brazil, Russia, South Korea, Indonesia, and others suffered terribly during the 1980s and '90s as the World Bank and IMF lent money on the condition that these harsh and inhumane policies be implemented.[62]

Thankfully, many of these countries, including Brazil, Chile, Argentina, and South Korea, are now disengaging themselves from these strictures and are separating themselves from these harsh institutions, caring for their people and their societies in humane ways. Brazil has refused to renew its agreement with the IMF, Nicaragua wants to leave its fund, Venezuela has split from both the World Bank and IMF, and Argentina is also splitting. This trend is also apparent in other countries, and the IMF global portfolio has been reduced from $81 billion to $11.8 billion in just three years. The World Bank apparently is equally unpopular with Ecuador and Bolivia, and they are no longer apt to toe its line. Finally, developing countries seem to have enough confidence to get out from under the perverse and dangerous control of the transnationals—the international banks and their invidiously dangerous influence and policies, which damage people and societies.[63]

Women tend to bear the brunt of these IMF policies, spending more and more of their day digging in the fields by hand to increase the production of luxury crops, with no machinery or modern equipment. It becomes their lot to help reduce the foreign debt, even though they never benefited from the loans in the first place. Their health suffers, and they become tired and anemic because of poor diets. Consequently, maternal and infant mortality is rising in many of these countries.[64] The price of food, too, is increasing, and wage cuts imposed by the IMF are severe. Millions of people can no longer afford to adequately feed, clothe, or educate their children in the Third World. One-fifth of the agricultural products grown there are exported to the First World.[65]

Meanwhile, as world markets become flooded with these "commodity" or luxury exports, the prices drop and the First World benefits from cheap luxury imports. Thirteen percent of the food eaten by Americans is imported. The European Union, Canada, and Mexico are the top food exporters to the United States. However, China is quickly catching up, currently exporting 3.3 percent of its food to the United States. Of the food imported from other countries into the United States, 98.7 percent is not inspected by the FDA.[66]

Over the last fifty years, Africa has received $568 billion in aid from the world, but the aid has done little to lift the continent out of poverty. What the people of Africa really need is a set of tools so that they can be educated. In addition, in 1988, seventy out of a hundred Africans were labeled destitute, with an annual per capita income of just $50–$115,[67] and the vast majority of people living on much less than this.[68] In most African nations today, the GDP is less than $200 billion per year. The First World determines the price of Third World–produced luxury goods, while the Third World has no say![69]

Most of the profits from commodity sales in the Third World go to retailers, middlemen, and shareholders in the First World. Only 15 percent of the $200 billion in the annual sales of these commodities in rich countries winds up in the countries that grow them. For instance, just 10 percent of the income from the sale of bananas is paid to the producing country.[70] In 2002, Venezuela produced 5.8 million short tons of bananas (1 short ton = 2,000 pounds), and almost all of it was exported to Europe and the eastern United States.[71] The workers in the Third World in general receive only 1–2 percent of the value that they create. Nicaraguan coffee pickers are paid only eighty cents (15 cordobas) for each bucket (29 pounds, or 13 kilograms) of coffee picked, plus food.[72] They receive 0.5 percent of the wholesale value of the coffee, while governments and rich plantation owners make $750 million per year. Coffee pickers in Ethiopia earn less than a dollar a day, but Oxfam recently announced that it will be implementing a program with the Ethiopian government that trademarks its coffee, which would allow the coffee industry to earn about $88 million a year—a plan that has the potential for farmers to earn more money than that paid for organic coffee or fair-trade coffee, a scheme that presently pays the farmers more for their produce than the poor compensation they receive from big multinational corporations.[73] In 2008, the average tea picker in India earned $1 a day working twelve hours a day seven days a week.[74] In 1985 in the Caribbean, the people actually starved beside fields growing flowers and tomatoes for export to the United States, which was once their property. Conditions have improved somewhat, but large corporations run a lucrative flower industry, and large quantities of pesticides are used liberally on the crops.[75]

Wealthy countries impose tariffs or trade barriers on processed goods, but none on raw materials, thus ensuring that poor countries remain in poverty. For instance, in 1985, British tariffs on raw cotton were zero; on cotton yarn, 8 percent; and on cotton T-shirts, 17 percent.[76] So the Third World can never break the poverty cycle, because First World tariffs work against the importation of manufactured goods from the Third World. A Third World country is defined as one that exports raw materials and imports finished goods. But processed goods are worth much more money than raw materials are.

Thus the spiral continues: increased debt leads to more cash crops and environmental degradation, which leads to flooded markets in the First World and lower prices, with decreased return to the Third World. Therefore, the debt increases in the developing world, and the result is malnutrition, starvation, and helplessness.

To make matters worse, the IMF also decreed that governments of debtor nations must severely curtail spending on social programs, as already described. Unemployment in Mexico is at a relatively low 3.2 percent at the moment, but underemployment in 2006 was roughly 25 percent.[77] Because of NAFTA (the North

American Free Trade Agreement), Mexico decreased its capped price on tortillas, which increased US corporate profit margins but also caused the price of tortillas to double. Many poor Mexicans rely on tortillas for half their daily calorie intake.[78] About 40 percent of the money that Mexico earns from exports goes to service its debt—the short-term debt has actually declined, but the overall debt has increased 50 percent since Mexico joined NAFTA,[79] and real wages have fallen since 2001.[80] A decade after NAFTA went into effect, Mexico's real wages are lower than before the agreement. Inequality and poverty have grown.[81]

Huge US-based corporations called "agribusinesses" also grow food in Third World countries for foreign markets—mainly the American market—and the debt crisis ensures them a plentiful supply of cheap, if not slave, labor and cheap land. They pay virtually no taxes to the host country. Ninety percent of the protein fed to British animals comes from underdeveloped countries. Sixty percent of dietary protein for humans in developed countries comes from animal products, but that figure is only 22 percent in developing countries. The meat consumption of people in the First World has increased 5–6 percent annually in the past twelve years, but the grain production remains static, even though the demand to feed people increases daily. Eight hundred million people in the world suffer from chronic malnutrition—17 percent in developing countries, 34 percent of which are in sub-Saharan Africa.[82] Because of the massive rise in food prices secondary to the rise in oil prices and the diversion of food to biofuels, a hundred million people in the world are now hungry. This is a global emergency.

Tanzania, which organized the best health care system in Africa in the 1970s, experienced a two-thirds cut in real wages, and the health care system rapidly declined. In the 1990s, the Tanzanian real GDP grew moderately, at a rate of 3.7 percent per year, but by 2005 it was growing by 6.8 percent annually. Despite the growth, however, the poverty level barely changed between 1991 and 2001. Today, 57.8 percent of the people live on $1 a day; 89.9 percent, on less than $2 a day.[83] Hospitals are collapsing, and maternal and infant mortality is on the rise. School buildings are falling apart, and many children no longer have the access to primary education that their predecessors had in the 1970s. This is a tragedy of monumental proportions. Even worse, this situation is typical of all the debtor nations.

By 2005, forty-four debtor countries were locked in huge repayments. Seven of them are responsible for nearly half the total Third World debt. The top poorest are Ethiopia, Ghana, Honduras, Niger, Rwanda, Nicaragua, and the Philippines. In 2005, some of the world's richest nations agreed to cancel $40 billion of debt, owed mostly by Africa. Britain agreed to pay $960 million and the United States up to $1.75 billion in compensation until 2015.[84] Many critics argue that this proposal benefits only a limited number of the poorest countries (eighteen in all). In addition, the debt would be canceled only for those who owe the World Bank and IMF, but not for those owing money to any Inter-American Development Bank (IDB) creditors. Latin American countries are still required to pay $3.3 billion to the IDB. As a whole, then, developing countries owe a massive $2.4 trillion.[85] Some observers say that the international banking system would be threatened if any defaulted on their payment. In 1980, the net funds transferred as loans from the First World to the Third World amounted to $40 billion, but in 1983 the situation was reversed, and the debt the Third World owed to the First was $20 billion greater than the loans it had received.[86] The Latin American countries together owe billions of dollars, and theirs are disastrously poor economies. How can they possibly pay this?[87]

In Britain, the banks were convinced on humanitarian grounds to put aside (forgive) half of the world debt and, in so doing, they were granted a huge tax relief. In fact, they made enough money from this strategy to immunize 450 million children against preventable diseases. They didn't, of course. From 1987 to 1989, using these maneuvers, British banks made $1.6 billion, which is more money than the British government paid to poor countries in foreign aid.[88] So the banks make money whichever option they choose—all on the backs of millions of starving, uneducated, sick children.

Third World debt does produce complications within industrialized nations other than the inconveniencing of the banks. Normally, poor countries buy and import manufactured goods from the First World, but in 1987, Latin America, instead of importing a surplus of $1.3 billion of goods as it had in 1980, incurred a $14.1 billion deficit. It could then no longer afford to buy products made in the First World. As a consequence, the United States lost two million manufacturing jobs. Meanwhile, ironically, the profits of the six largest commercial US banks rose by 60 percent between 1982 and 1986. So American workers suffered and the Third World suffered, but the banks flourished.[89] In 2001, Latin America had a $54 billion deficit, but by 2006 it had an estimated $35 billion surplus. A shift in the trade balance of nearly $100 billion in 2006 gave Latin America about 4 percentage points on GDP. In addition, the volume of remittances between 2001 and 2006—between $26 million and $55 million—helped significantly.

Though economic growth has occurred recently, Latin America continues to fall behind other regions of the world because investment is very low—21 percent, compared with Asia at 36 percent and Europe at 25 percent. Furthermore, the ratio of public debt to GDP did not decrease.[90] US bank profits from 1990 to 2006 were high: equities peaked in 1993 at almost 16 percent; equity profits in 2006 were about 13 percent. Assets peaked in 2003 at 1.4 percent; in 2006 they were at 1.3 percent. Gross income of all US banks in 1997 was $339 billion and by 2006 it was $551 billion.[91] So the banks continued to do well at that time, even though many of them are now crashing as a result of the subprime crisis. Then, ironically to shore up a failing economy, the federal government is stepping in to save some of these banks. So it is socialism for the banks but "capitalism" for the ordinary people, who are on their own.

Inflation in the debtor nations is horrendous. When I visited Rio de Janeiro in 1989, I could not quite grasp that new Brazilian banknotes had been printed three times that year. I never did come to understand the value of the different currencies, and only later did I discover that the inflation rate in 1989 had reached about 2,000 percent. I also learned that, behind the facade of great wealth in Rio and Brazil owned by 5 percent of the population, existed eighty-six million people suffering from malnutrition.

Because of new leadership, however, by 2007 Brazil's inflation rate was only 4.12 percent.[92] In 1992, the number of people in Brazil living on $2 a day or less was 36 percent, but by 2006 it was only 19 percent of the population. The gap between rich and poor that I had observed in 1989 had fallen between 2001 and 2006 by 5 percent. Brazil implemented a family stipend plan that gives about eleven million families (a quarter of the population) up to $50 a month for an indefinite period of time. The government spends almost $500 billion, about half its GDP, on social programs. Marcelo Neri of the Getulio Vargas Foundation, a Brazilian business school, said, "The 90s were the years of economic stabilization. This decade is going to be remembered as the era of falling inequality."[93]

The total amount of money owed by the Third World is $2.4 trillion, which is double the $1 trillion that the global community spends on weapons each year. Debt is not an adequate way of solving the problems of millions of people living in poverty, and death is not a way of solving conflicts. War is obviously obsolete in the nuclear age. If $1 trillion were transferred annually from the death industry to the life industry, 80 percent of the world could feed, clothe, house, and educate itself and provide excellent modern health care and contraceptive services, while the First World could concentrate on solving the problems of ozone depletion, the greenhouse effect, deforestation, global pollution, nuclear disarmament, and species extinction.

So simple—yet so difficult because of human nature. I stayed with some wealthy investment bankers recently at Palm Beach, Sydney, who told me the world would never change, because people are selfish. I suspect they were talking about themselves. It was Einstein who stated prophetically, "The splitting of the atom changed everything save man's mode of thinking, thus we drift towards unparalleled catastrophe."[94] "Man's mode of thinking"—but what about women and good, caring men who need also to change their way of thinking if we are to survive?

SOME ILLUSTRATIONS OF THIRD WORLD DEBT

Having described the generic problems of the Third World countries and First World banks and corporations, I want to outline some specific conditions that require our immediate attention.

In 1984–85, when thirty-five million people were dying from drought-induced starvation in Africa, Bob Geldof and others organized immense fund-raising "food aid" concerts, which saved the lives of millions. But even so, when the drought broke, nineteen million people remained malnourished, and by 1990 another drought had hit Africa. Again, climatic catastrophe has combined with ongoing tribal wars, and Third World debt has caused massive population displacements and starvation. The Intergovernmental Panel on Climate Change's (IPCC) "Fourth Assessment" (April 2007) states that Africa is the world's continent most vulnerable to climate change, which will have the greatest impact on its estimated 840 million people. The number of cases of malaria, the second biggest killer of young people after starvation/malnutrition, will greatly increase. The report states, "Drought-affected areas will likely increase in extent. Heavy precipitation events, which are very likely to increase in frequency, will augment flood risk." About 25 percent of Africans (or two hundred million) currently experience high water stress—insufficient availability of water. By the 2020s, between 75 million and 250 million Africans will experience lack of water; and by the 2050s, those numbers will have increased to between 350 million and 600 million. These figures are almost beyond imagination.[95]

It was thought that the 1990 drought in Africa may have been a result of the impending greenhouse effect. According to data at that time from NASA, the British Met ("meteorological") Office, and the US Commerce Department's National Oceanic and Atmospheric Administration (NOAA), 1990 was the warmest year on record, and six of the seven warmest years in the previous 140 years of recorded history had occurred since 1980.[96] These postulations were almost certainly correct. Over the past twenty years, eastern Africa has experienced an increase in dipole rainfall patterns, meaning that more rainfall occurs in the north, while the south experiences declining levels of rain. The El Niño–Southern

Oscillation (ENSO) has had a major effect on rainfall at inter-annual scales. Extreme floods due to ENSO occurred in 1997, yet one-third of Africans live in "drought-prone areas." Only 62 percent of the African population in 2000 had access to "relative water abundance."[97]

Clearly, greenhouse effect, drought, debt, and exploitation by the developed world have led to misery and chaos for millions. These African droughts are worse than earlier ones. Previously, crops failed and thousands of animals died, but now there is no water in many populated areas and whole villages have been uprooted, their people rendered nomadic.

In the 1960s, Africa was a net exporter of food, but over the last forty years the population in sub-Saharan Africa has grown at a rate of 4 percent per annum (the highest growth rate in the developing world). Thirty-seven percent of this population lives in shanty towns while agricultural production increases at a rate of 2 percent. Since 1962, per capita food production has decreased yearly. The sub-Saharan African population is expected to grow by 143 percent between 2000 and 2050.[98] According to the World Bank, if present trends continue in Africa, the food gap of 15 million tons in 1990 will increase to 200 million by 2020.[99] From 1960 to 2003, the United States and other industrial nations spent $2.3 trillion on bilateral and multilateral development assistance that was supposed to help developing nations gain economic growth, and about one-fourth of this money went to sub-Saharan Africa.

To sum up the world food situation, in the years 1950–84, world grain production grew faster than the population because of the introduction of new breeds of rice and wheat, which could yield two or three crops a year, and the more intensive use of fertilizers and pesticides. Food production rose 2.6 times, increasing by 40 percent the amount available for each person on Earth. In China, the per capita increase was 80 percent; in western Europe, 130 percent.[100] In 1950, China's grain harvest was 90 million tons, and by 1998 it was 392 million tons. But five years later, the harvest had dropped to 322 million tons. Wheat, rice, and corn stocks continue to fall as prices increase because of water shortages in the north. Chinese farmers are also losing land because the deserts are expanding; and urban, industrial, and highway construction is stealing their land.[101]

Globally, since 1984, there was a reverse trend from the highs of the 1950s to the mid-1980s, as grain production decreased each year, declining 14 percent from 1988 to 1991. This decline was caused in part by the serious 1987 drought in India and by the 1988 droughts in Canada, China, and the United States. Other causes for the decline were erosion of topsoil, desertification, salinization, and waterlogging resulting from irrigation, depletion of groundwater supplies, and diversion of irrigation water to nonagricultural uses.

By 2006, world grain harvest output was just under 2 billion tons, down 24 million tons from 2005. Water reserves are shrinking, while temperatures and populations increase. In Africa, famines and drought have struck thirty-four countries and over thirty million Africans.[102] (Desertification affects the livelihood of more than one billion people globally, and 40 percent of the Earth's surface,[103] while each year about 52 million acres [21 million hectares]—an area equaling the size of Kansas—are destroyed when farming lands are converted to deserts because of agricultural mismanagement.)[104] Another significant factor affecting world grain shortage is the rising income of millions of consumers worldwide, who are eating higher on the food chain, consuming more poultry, beef, milk, eggs, and farmed fish. For instance, world meat production in 1950 was 44 million tons, and by 2005 it had grown to 265 million tons to meet the growing demand.[105]

The world food situation is very confusing because, on the one hand, there are millions of starving and malnourished human beings and, on the other, First World countries are producing enormous quantities of "excess" food, which is either being stored in silos and underground caves or being dumped into the sea, or unloaded onto Third World markets at very low prices. In 1969, the European countries, Japan, and the United States subsidized their farmers by $25 billion yearly. Now, taxpayers in Europe, Japan, and the United States spend about $1 billion a day subsidizing their farmers.[106] In the early 1990s, a time when the world butter price was $300 per ton, Europe paid its farmers $1,400 per ton for butter.[107]

In 2007, the world price of butter was estimated to be between $2,275 and $2,391 per ton,[108] and the European Union offered farmers export subsidies of 136 percent of the international price of butter.[109] From 1995 to 2005, US farm subsidies alone totaled over $1.6 trillion.[110]

These cynical, vote-buying policies have a threefold effect:

1. They ensure that the huge farming lobby will continue to support the existing governments in the United States and Europe during elections.
2. They lead to the production of huge quantities of artificially cheap subsidized food.
3. Third World countries rely more and more on this cheap imported food because most of their land is now used to grow cash crops for developed countries to pay back their foreign debt.

Bananas, cotton, rubber, cocoa, sugar, coffee, and flowers are exported at low prices because of a world glut in the wake of shortsighted IMF policies that forced many debtor nations to grow food and crops for export. As a consequence, poor peasant farmers are starving or forced to buy imported food because they have so little land for growing food. In addition, the price of locally grown food suffers when aid in the form of food is given cheaply or for free.

To summarize, in 1980 about $35 billion was transferred from the poor Southern Hemisphere to the rich Northern Hemisphere because of adverse trade terms, repatriation of profits, "brain drain" (migration of well-educated people), and interest payments. OECD (Organisation for Economic Co-operation and Development) countries spend over $75 billion a year on subsidies to their farmers, thereby forcing their consumers to pay more than $240 billion extra for food per year. That's about six times more than the developed countries provide the developing countries in aid per year. Halting OECD farm subsidies therefore could transfer $40 billion in export revenues to poorer countries and about $3 billion to Africa.[111] In fact, the real aid passes from the poor southern countries to the rich northern ones. The overall relation between population increase and food production appears grim: more people and less food.[112] The remedy lies only in the application of money, science, and political will by the rich and poor countries in concert. No person should starve in this world; conversely, no person has a right to be rich on this planet while others starve. There is potentially enough food for everyone, and each human life—be it African, Indian, Asian, or European—is as precious and sacred as the next.

Recently, however, another problem has arisen that is affecting world food supplies. According to a confidential report prepared for the World Bank, biofuel production is forcing global food prices to rise by 75 percent. Rising food prices have already pushed a hundred million people below the poverty line and have sparked food riots in Egypt, Bangladesh, Mexico, Morocco, Yemen, Uzbekistan, Guinea, Mauritania, and Senegal. The report states that "without the increase in bio-fuels, global wheat and maize stocks would not have declined appreciably and price increases due to other factors would have been moderate." Farmers have been encouraged to set aside land for biofuel production, and this situation has sparked financial speculation in grains, which has driven prices higher. More than one-third of US corn now goes to biofuel for cars and not to feed people.[113]

General Agreement on Tariffs and Trade

The Third World has suffered yet another insult to its basic integrity, at the hands of the transnational corporations of the United States and other rich nations. These corporations straddle the oceans and conduct their affairs and business in many, and sometimes most, countries of the world. The term "transnational" therefore seems more appropriate than "multinational." Companies that began in, say, the United States grew and became *multinationals* as they extended their influence into several other countries, but they became *transnationals* when they moved into almost every country of the world. Transnationals tend to be headquartered mainly in the United States, the United Kingdom, France, Belgium, West Germany, Japan, and Switzerland. Two of the more striking examples are Coca-Cola and PepsiCo. Transnationals, in practice, are impersonal, faceless organizations answerable to no one except their shareholders. Many of them have budgets larger than the GDPs of most countries. They seem to be bound by no moral principles, guided only by the profit motive.

Transnationals have enormous power to control and manipulate the population, consumers, and politicians. They have their own global intelligence and communication networks, they can afford to buy business favors (bribes) from politicians, and they run million-dollar public relations campaigns. And they can pay for the very best lawyers to counter any legal challenge.[114]

Hundreds of years ago, before corporations were invented, poor countries were invaded and colonized by richer, more powerful countries. The people in the poor countries were subdued and controlled by armies, and their natural wealth in spices, gold, silk, opium, forests, and minerals was plundered and stolen by their wealthy occupiers. Millions were killed and slaves were taken. Colonialism was often supplemented by gunboat diplomacy, and more recently, covert operations have been used to subdue difficult Third World countries. Since 1945, the United States has intervened militarily in foreign countries on the average of once every eighteen months, primarily in order to control its markets and provide access to cheap labor or natural resources.[115] Since President Harry S. Truman, the United States has intervened militarily abroad fifty times without the approval of the UN Security Council.[116] Of course, the most recent and tragic example of these policies is the war in Iraq, which was clearly instigated so that private US companies could regain access to Iraqi oil but killed over one million Iraqi civilians in the process.

Despite past repression, though, a global movement toward nation-state autonomy is gaining ground. After India achieved its independence from England, other colonial countries followed a similar path to freedom in Africa, Asia, and Latin America. Yet, for all their immense influence in many countries, transnational corporations are still not satisfied. They feel threatened by the rising nationalism and independence of many Third World countries, and they initiated a brilliant round of negotiations aimed at bringing all other countries under their ultimate control. If these words sound too strong, let me elaborate.

The organization known as the General Agreement on Tariffs and Trade (GATT), established during the reconstruction period after the Second World War, was chosen by the US government and the transnationals as their vehicle to regain total sovereignty over most nations in the world. A round of talks under GATT auspices was initiated in Uruguay in 1986.[117]

This round of GATT talks was considered by many transnationals to be necessary for reorganizing the international economy and solving economic problems into the next century. All relevant meetings were conducted behind closed doors in absolute secrecy, and only the major industrialized nations of Europe, Japan, and the United States, along with representatives from the transnationals, were invited to attend. The latter were also present as "advisers" to their national delegations. This decision-making process was called the "green room consultations," after the wallpaper of the GATT director general's conference room in Geneva. GATT documents and decisions were restricted and secretive, and the media were barred. All other countries, including the developing nations, were excluded from the talks, as were nongovernmental organizations (NGOs), which represent church groups and humanitarian interests.[118]

The basic mission of the 1986 GATT talks was to enhance the influence, power, and control of transnationals in the Third World. It would curb the right of governments to use their economies and tax dollars for the benefit of their people, obliging them, instead, to provide "space" for the transnationals. Any transnational commencing certain operations in a specific country would be free to expand its activities into any other area (say, from mining to agriculture). The property rights of these rich foreigners would take precedence over those of the nationals. The transnationals would be allowed to produce or import goods or services as they pleased and to decide which technology, if any, could be imported and used in a certain country. In other words, countries would become prostituted to the transnationals.[119]

These GATT regulations were designed to reshape the existing international trade system to give transnationals total freedom to operate globally. These US-inspired efforts succeeded in forcing developing nations to decrease or eliminate the regulation of investments and the control of the operations of foreign companies on their soil. In many countries now, these corporations effectively control mining, manufacturing, banking, transport, insurance, media, communications, advertising, wholesale and retail trade, auditing, and legal practice. The specific governments have also been forced to introduce laws protecting and enhancing patents owned by the transnationals and other industrial rights. Because of these new international laws, people in the Third World have discovered that they do not even own the rare plants and seeds that were discovered in their forests and that could be put to commercial use. Farmers often cannot store seed for the next season or even breed their own cattle, because the transnationals own the seed and the livestock, and the farmers are obliged to buy seed and livestock from those companies.[120]

At present, many universities in rich countries work hand in hand with corporations. Although the US government sponsors or pays for most research and development, corporations benefit from much of the resulting knowledge. Since the Reagan years, almost all new scientific knowledge (even some medical) has been classified because it has relevance to the military-industrial complex or because of corporate ownership or jealousy. This dynamic is called "commercial in-confidence." Thus, much scientific research or knowledge is now classified and/or owned by corporations—a situation that violates the ethics and principles upon which science is based. Fundamentally, the world's knowledge and technology base is the private property of the major transnationals.

Research and development (R&D) conducted by approximately three million corporate-related scientists and engineers cost more than $226 billion in 2005.[121] In 1990 almost 70 percent of the world's scientists and approximately 85 percent of the world's expenditures were devoted to R&D. General Motors spent more were devoted to R&D than did the country of Belgium, and IBM more than doubled the amount spent by India, with its one billion inhabitants. Transnationals hold the bulk of the world's patents, over 90 percent of which originate in developed, capitalist countries.[122]

The transnationals are the world's major sources of technology, and they control it legally through patents, investments, cross-licensing, and services. At the moment, a huge chasm separates the knowledge-rich northern and the knowledge-poor southern countries, and it is about to widen, as transnationals jealously guard and protect their information, which, under no circumstances, will they share with Third World countries. Thus, these handicapped nations will never develop or improve their standard of living and will remain poor and beholden to those in the Northern Hemisphere and the First World.[123]

At the same time, though, old redundant technologies are exported to the poor south, while the domestic markets of the rich north are closed to most manufactured or industrialized goods from the south. Toxic drugs banned by the FDA, poisons, pesticides, herbicides, and other noxious chemicals are freely dumped by transnationals into the Third World. In 2005, world pesticide sales totaled $32.5 billion.[124] In one instance, although the consumption of pesticides recently decreased, problems related to unregulated use are quickly rising—poisoning of workers and their families who are exposed to the toxic fumes as the chemicals are sprayed on the crops. (Some half a million people are poisoned each year by pesticides, mostly in the Third World.) In India, most pesticides are used for cotton (45 percent), followed by rice (22 percent).[125] The GATT agreement allows foreign corporations even more access to developing countries and forces countries to accept the foreign agricultural products flooding their markets.[126] Compliance is compelled by the threat of trade retaliation. The transnationals are able to take and sell what they want while restricting access of the developing world's goods to the rich, developed world.[127]

Great advantages have been reaped by agribusiness transnationals, which have taken over enormous tracts of Third World land to grow cash crops, while peasants suffer and forests are destroyed. Of course, access to all this land gives them ample opportunities to use vast amounts of toxic pesticides and artificial fertilizers to grow food. In 1991, three-fourths of the fruits and vegetables consumed in the United States came from the Third World.[128] From 1994 to 2004, fruit and vegetable imports to the United States doubled in value, to a total of $12.7 billion annually.[129] Of course, this international commerce in food adds enormously to global warming, as all this produce travels thousands of miles before it lands on the table. Incidentally, for a historical perspective, these are the same agribusiness companies that took over the land when thousands of proud Iowan farmers went bankrupt during the agricultural recession induced by Reagan policies.

Under the GATT agreements, Third World countries lost their ability to control timber exports. In the meantime, fast-food hamburger chains urged the US government to use "free trade" to abolish US beef import quotas or tariffs—an action that has tragically led to increased destruction of the Amazon forest for grazing land.[130]

Let me give an example of GATT's power, control, and philosophy. In August 1991, a tribunal of trade experts from Hungary, Uruguay, and Switzerland mandated that GATT prohibit its 108 members from imposing import restrictions founded on environmental issues. This policy overthrew the US ban on tuna from Venezuela, Mexico, and Vanuatu. Tuna from these and other countries, including Costa Rica, France, Italy, Japan, and Panama, was banned from the United States because their fishing fleets killed many dolphins during the tuna catch. However, some US companies still abide by the dolphin-safe tuna policy. GATT has had the ability to overturn many essential national environmental laws in the name of free trade. The US Congress "fast-tracked" legislation—and quickly endorsed all of the GATT proposals as they appeared—an extremely dangerous development.[131]

Thus, the term "free trade," which the White House likes to bandy about, means free trade for the transnationals and "total trade restriction" for all countries opposed to the global exploitation by corporations. Another favorite phrase describing GATT objectives is "level playing field," which really means that the transnationals can exploit other countries with impunity.

GATT has affected national sovereignty, the environment, Third World people, farming, food and safety standards, and ownership of minerals and national resources. In other words, GATT has achieved through a new legal world order what colonialism, gunboat diplomacy, and covert operations could not attain—total transnational control of the resources of all countries. Because transnational corporations will control many domestic services, including the media, the only role left for Third World governments will be to maintain law and order and control labor and unions.[132] We can witness these changes in Australia alone, where News Limited owns 75 percent of the local print media nationally and many of our treasured publicly owned institutions—including Qantas, electricity-generating services, and the once publicly owned telephone company—have been sold off to private multinational corporations.

The Uruguay rounds of negotiations from 1986 to 1994 formed the World Trade Organization (WTO),[133] which replaced GATT. The intention was to reduce specific tariffs and to set new rules for trade. By December 2007, 151 country members were signatories. However, these WTO negotiations have had a rocky course. Because the developing world was opened to exploitation by the transnationals and the advanced industrial nations set the agenda that opened markets for goods and services to their comparative advantage, the next meeting, in Seattle in 1999, was a wild confrontation between protesters from around the world and hundreds of police. Thousands of protesters converged on the scene, and the meeting was a disaster.[134] The next meeting was held in Doha in Qatar in 2001, far from any demonstrators, but the developing countries were skeptical that they would be treated fairly, and the Doha round failed.

In 2008, the United States was accused of not adhering to the WTO treaty regarding $15.5 billion of Internet gambling.[135] Another issue that has been problematic for the WTO since the 2001 Doha round of talks is how exactly to induce rich countries to reduce their domestic farm subsidies, which distort the prices of common trade products, such as cotton, sugar, and corn, to the detriment of developing countries.[136] The US Congress, for example, passed a farm bill in 2002 that doubled its already substantial domestic farm subsidies.[137] This question was answered in July 2008 when the WTO met in Geneva. The developing countries were close to accepting a deal that included modest reductions in agricultural protections for rich countries and modest cuts in industrial tariffs for developing countries. It also contained a clause for a "special safeguard mechanism" allowing developing countries to raise their tariffs if unexpectedly large increases of food from subsidized farmers in rich countries were to swamp local markets and further threaten poor farmers. In the end, the United States steadfastly refused to accept this clause, and the talks collapsed in a bitter split between developed and developing countries, particularly China and India.[138]

Perhaps this is what President George H. W. Bush meant in his 1991 State of the Union address when he talked about "a new world order." China is another vast market in the making. On my visit in 1987, I came down to breakfast in the Great Wall Hotel in Beijing, built by a US corporation. I already felt uncomfortable in this atmosphere of affluence, for the hotel towered over thousands of dirt-floor hovels. I encountered an American in the lobby and asked him uneasily how he felt about China. He settled back in his chair with one leg draped over the armrest and said, "Oh, there are great opportunities here." I nearly vomited. I had spent three weeks traveling around China, developing enormous respect for its industrious people and their ancient civilization. They had been able to extricate themselves from the clutches of British colonialism over the preceding thirty years; had managed to feed, educate, house, and clothe one billion people; and had successfully dealt with their overpopulation problem. The revolution had some severe setbacks, as do all political systems, including the cultural revolution of the 1960s, with loss of civil liberties, and, in 1989, the Beijing massacres. But what will happen to this revolution if the transnationals take over? They will almost surely bring the end of equality and the beginning of exploitation, divisions between rich and poor, and a reversion to colonialist times. (I wrote the previous sentence in 1991, and indeed my predictions are coming to pass.)

Most people are aware of who holds the purse strings. We know who controls the US government. During my talks to hundreds of thousands of Americans, I always ask, "Hands up, those of you who believe you live in a true democracy." One out of two thousand people may raise a hand. I then ask, "Who runs your country?" and they answer, "The corporations do."

9 The Manufacture of Consent

In order to appreciate the influence and power that the transnational corporations exercise in the Third World, we must understand how they obtained this power in the first instance. Many of the global transnationals had their origins in the United States as small businesses over the last century. By a combination of competition, hard work, takeovers of smaller companies, and some ruthlessness, they became enormously wealthy and powerful.

How did they manage to gain the support and cooperation of the American people as they achieved their power? The truth is that they have not only been competitive and somewhat ruthless in the Third World but have also exhibited a history of exploitation and subtle coercion within their country of origin. Most Americans are not familiar with this history. Although this "manufacture of consent," to use the phrase coined by the respected journalist Walter Lippmann, was investigated and soundly denounced by national church groups and congressional committees, the political and propaganda activities of US corporations have continued unabated over the years and to this day remain largely unknown.

The story of corporate propaganda explains, I believe, the

strange and powerful patriotism and nationalism of the American people, their ready acceptance of propaganda and media manipulation, and the reality that US workers are not represented by a broad-based, powerful union movement. It also explains why the minimum wage in 1991 was only $4.75 per hour and why there are insufficient occupational health and safety standards, no uniform free health care system, and no national system of free higher education. The minimum wage had increased in California in 2008 to $8 per hour, and in New York in January 2007 to $7.15—a paltry rise, considering the cost of living increases since the early 1990s. Finally, the federal government raised the minimum wage on July 24, 2008, to the princely sum of $6.55, and one year later it is set to increase to $7.25 an hour.[1]

By comparison, the people of my native country are quite blasé about nationalism or patriotism; Australians rarely fly flags. But the minimum wage is $13.19 per hour (converted to US), and the union movement is now relatively strong. Workers have excellent occupational health and safety standards, and we have a free health care system. University education was once free but now must be paid for because the influence of the transnational philosophy and "economic rationalism" has infiltrated Australia and its politicians. Still, it is not nearly as expensive as a US college education.

This chapter and the next will continue to examine the etiology of the global ecological crisis. This may be a difficult chapter for some people to read and digest because it challenges many basic beliefs of American society, but I hope you have the courage to look at the truth.

I rely in the following discussion on the work done by the late Alex Carey, an Australian psychologist who lived in the United States during the 1970s and '80s. He was fascinated by the origins of American culture and made an extensive study of the history of propaganda and its political effects.

From the beginning of the twentieth century, large-scale professional propaganda campaigns have been waged by American business in order to shape public attitudes to accept and endorse the capitalist system, and this propaganda has changed the direction of American society.[2]

The campaign began between 1880 and 1920 in Britain and the United States, when the right to vote was extended from 15 percent of the adult population to 50 percent. This popular franchise immediately posed a threat to the rich minority, because as real democracy was instituted, people would naturally be voting for laws that supported their own health, education, and welfare. For the first time, their tax dollars would be used to support the majority of the population and not just the rich. In 1909, two leading scholars—Abbott Lawrence Lowell (president of Harvard) and Graham Wallas (a leading British student of democracy)—warned that the consequences of those new laws might be dangerous. They said, "Popular election may work fairly well as long as those questions are not raised which cause the holders of wealth and power to make full use of their resources. If they do so, there is much skill to be bought, and the art of using skill for the production of emotion and opinion has so advanced, that the whole condition of political contests would be changed for the future."[3] In other words, if the power of the rich is challenged, they will use their money to intimidate and coerce people, to buy votes, and to produce a mandate for themselves.

By 1913, a congressional committee had been established to investigate the activities of an organization called the National Association of Manufacturers (NAM), which represented many US businesses and had already begun disseminating vast quantities of literature with the apparent intention of "controlling" public opinion in the fledgling democracy. But public opinion moved away from and did not support corporate philosophy during the First World War, which ended in 1918, because the American people were encouraged to work together for the good of the country. Women took equal jobs side by side with men, and a mood of unselfishness and generosity prevailed in the country.

Propaganda Techniques

The techniques of propaganda were developed during the First World War when the American people were reluctant to become involved. President Woodrow Wilson and others initiated a large and very effective campaign, under the supervision of the Committee on Public Information, headed by George Creel—a Denver newsman—to convince the nation that Germany, whose people Creel called Huns, was the seat of all evil.[4]

Propaganda is "the organised spreading of ideas, information or rumour designed to promote or damage an institution, movement or person."[5] World War I marked the first time in history that propaganda had been successfully conducted on a large scale. Within six months, the American people were devoted to hating the Germans and to defeating them in the war effort. (Does this sound familiar? Replace Germany with Iraq, and we might as well be talking about the present day.) Public opinion at that time had been so aroused that grotesque campaigns of witch-hunting and Americanism abounded.[6]

One of the creators of propaganda during the war was Edward Bernays, a nephew of Sigmund Freud, whom he closely resembled. (I met Bernays in 1981 on a sultry, rainy Boston day, when he offered to help in my anti–nuclear war campaign. In the end, he was unable to contribute, but I learned some rather remarkable facts. As we sat drinking tea, looking over a narrow Cambridge street, Bernays told me proudly that he was the person who had taught women to smoke, by dressing them in beautiful green gowns, placing a cigarette in their hand, and adorning *Vogue* magazine with their photographs. I felt ill.) Bernays headed the transfer of wartime propaganda skills to the business arena. When the war ended, he wrote, business "realized that the great public could now be harnessed to their cause as it had been harnessed during the war to the national cause, and the same methods would do the job."[7]

Incidentally, in 1919, Samuel Instull, who headed a vast utilities empire, took over the entire hierarchy of the "information committees" set up by President Wilson during the war, and he used them to great advantage to protect private utilities against the threat of public regulation or ownership. Here lie the seeds of the public's remarkable support over the years for private utilities and for nuclear power.

After the war, propaganda was recognized as a tool that corporate America could use to further its own agenda. In 1919, some 350,000 US steelworkers went on strike and demanded shorter working hours and higher pay. (They were working eighty-four hours per week.) Because they had felt proud of their work during the war, the steelworkers naturally expected to be well treated. But they were not. Instead, the United States Steel Corporation (U.S. Steel) bought full-page advertisements in newspapers to encourage the strikers to return to work and to accuse the strike leaders of being Bolsheviks and Reds (at a time when the Russian Revolution was in its infancy) and of being Huns. They also told the American people that the price of steel would soar if the workers prevailed. How could the US corporations call proud American workers communists and Germans when they had rallied so staunchly behind the war effort?

Unfortunately, a gullible public was largely persuaded by this propaganda offensive. The strike was defeated by the engineering of public opinion, which had been successfully turned against the workers. By the time the strike ended, twenty workers had been killed, the hours and wages remained the same, and the price of steel had gone up. So the steel industry won.

Organized unions, I believe, are the best and only vehicles for the representation of the true interests of the working people of the United States—health care, occupational and safety standards, wages, working hours, and so on. They constitute the sole force that can take on, and to some extent control, corporate power. But since 1919, US corporations have systematically worn down, demoralized, and destroyed organized labor by using the techniques of propaganda, Red-baiting, and intimidation.

Public Opinion

The successful propaganda campaign that ended the steel strike was then extended to American public opinion at large. Corporate America started the rumor that US workers and their leaders wanted to overthrow the federal government. It introduced this unsubstantiated notion through a public relations campaign in the media that led to an intense period of virulent anticommunism, in the years 1919–21. The witch hunts and blatant Americanism also continued after the war, and people were jailed for practicing their right of free speech.[8] As a result, many American citizens felt persecuted and alienated within their own country. This propaganda campaign proved very effective, and the American public was persuaded to support the rights of rich citizens and corporate power, while support for civil liberties, social reform, and the labor movement declined.

In the 1920s, Edward Bernays called this use of propaganda "the engineering of consent,"[9] and Harold Lasswell, for fifty years the leading US scholar of propaganda, said in 1939 that propaganda had become the principal method of social control. Lasswell remarked, "If the mass will be free of the chains of iron, it must accept the chains of silver. If it will not love, honour and obey, it must not expect to escape seduction."[10] In other words, if ordinary people gain power in a democracy through the vote, then the rich will find another way to maintain control.

Everything went well for corporate America during the 1920s, but the country suffered during the Great Depression of the 1930s. Tens of millions of people lost their jobs, the banks collapsed, and people starved. Among the poor and indigent arose a great wave of hostility and animosity directed toward big business and corporate power, and people demanded a more equitable distribution of wealth.

In 1932, Franklin Delano Roosevelt was elected president, and his administration launched the New Deal, which cared for and gave succor to millions of unemployed, depressed people. Mass employment schemes were initiated to rebuild cities, bridges, and roads, as were other public works. The common people of the United States loved and trusted FDR, being mesmerized by his "fireside chats," broadcast on the radio. It became morally and politically acceptable to advocate government ownership, government programs, and socialism as such.

Businesspeople, however, never liked or accepted President Roosevelt, because they had temporarily lost the public's loyalty as a result of his leadership, so they set out once again to recapture the minds of the people. They spent millions of tax-deductible dollars on public "education" programs and on polls. They also taught "human relations" to their own workers in order to control their thinking.

In 1935, the by then renowned National Association of Manufacturers organized another massive propaganda campaign. The NAM's president told business leaders in 1935, "This is not a hit or miss program. [It is] skillfully integrated . . . to . . . blanket every media . . . It pounds its message home."[11]

In 1939, the La Follette Civil Liberties Committee of the US Senate reported that the NAM had blanketed the country with propaganda that relied on secrecy and deception. The NAM employed radio speeches, news cartoons, editorials, advertising, motion pictures, and many other techniques that did not disclose its sponsorship. One business-sponsored agency distributed a steady supply of canned, ready-to-print editorials to twelve thousand local newspapers, and some 2.5 million column-inches of this material was published.[12]

Polling

By the late 1930s, public opinion polling had been invented, and it turned out to be highly useful to business. It was employed, according to Alex Carey, as an "opinion sensitive radar beam," which continually assessed ideological drift in the population. The polling data were used by the industrial propaganda institutions to provide continual flow of data and feedback, so that they could define and redefine their pro-business messages to make them more effective. It was also used to evaluate public response to product marketing.

Nationalism

In 1945, the corporations invented a new method to sell their capitalistic philosophy, which they called "techniques for community ideas." They discovered that the American people were not very excited by the rather sterile concepts of capitalism or free enterprise but that they did exhibit a rather positive emotional response to the notion of "Americanism." From this new information, the corporations devised a formula that tied many fundamental values together:

> free enterprise = freedom = democracy = family = Christianity = nationalism = God.

The equal and opposite formula they devised went something like this:

> egalitarianism = equality = government interference = socialism = unions = communism = Satan.

These two formulas became the backbone of corporate philosophy and profit-oriented activities and propaganda, and they have been used ever since with undiminished success.

Just watch TV carefully for one night, and you will understand what I mean. A big, tough, hairy guy hefts a can of beer, the ad implying that all red-blooded Americans drink this brand of beer. The scene fades with patriotic music in the background. There may even be a suggestion of an US flag waving in the background. Advertisements also imply that freedom and democracy are God-given, that the family is the fundamental unit of American society, and that all is well with the world, if you buy the products in the ads.

Strikebreaking and Public Relations

In 1937, a steel strike erupted in Johnstown, Pennsylvania, when Bethlehem Steel refused to acknowledge the steel union. At that time, the corporations needed to gain control of a restive population, after the years of the Depression and the New Deal. So the local chamber of commerce joined with the NAM and Bethlehem Steel to orchestrate another propaganda campaign, using the steel strike as the fulcrum. The National Citizens Committee was organized and launched by local businessmen; it engaged an advertising agency and a public relations council. The committee broadcast its antistrike messages of "Americanism" twice over a national network, and two full-page ads appeared in thirty newspapers in thirteen states. The campaign was once again successful.

At the end of the strike, James Rand, of the Remington Rand Corporation, proudly announced, "Two million businessmen had been looking for a formula like this, and business had hoped for, dreamed of and prayed for such an example as you have set."[13] This antistrike tactic was called the "Mohawk Valley formula," and since that time this strikebreaking technique has been used in every major strike in the United States.

The Senate-based La Follette committee criticized the propaganda tactics of the NAM in the 1930s, building up to the 1940s, in the following way: "The leaders of the association resorted to 'education' as they had in 1919–21. They asked not what the weaknesses and abuses of the economic structure had been, and how they could be corrected, but instead paid millions to tell the public that nothing was wrong and that grave dangers lurked in the proposed remedies."[14] But the NAM continued to use its propaganda campaigns in the fight against labor unions. The corporations fought the most important strikes in 1945–46 in the press and over the radio, not the picket lines. In effect, business owners bypassed the workers and went over their heads to appeal to the public, using false and unfair statements.

Fundamentally, these corporate campaigns were designed to achieve three objectives: (1) to minimize wage rises and to maximize profits; (2) to oppose decent working hours, a minimum wage, occupational health and safety standards, and employee health coverage; and (3) to prevent government regulations from interfering with their activities.

In 1946, the US Chamber of Commerce distributed a million copies of a fifty-page article entitled "Communism in the United States." In 1947, similar distribution occurred for a pamphlet entitled "Communists within the Government," which alleged that about four hundred communists held important positions in the government. The most effective propaganda weapon for both employees and college students was found to be a comic book. The *National Association of Manufacturers News* of February 1951 proudly proclaimed, "If all NAM produced pamphlets ordered for distribution to employees, students and community leaders in 1950 had been stacked one on top of the other, they would have reached nearly four miles into the sky—the height of sixteen Empire State Buildings, a record distribution of 7,839,039 copies."

The American Advertising Council, which was established fourteen days before the United States entered World War II to combat people's enthusiasm about the New Deal and their disaffection with the free-enterprise system, represents large corporations and advertising agencies. In 1947, it announced a twelve-month, $100 million campaign, one of numerous related campaigns to "sell" the US economic system to the American people. Daniel Bell, a professor of sociology at Harvard, said in 1954, "The output is staggering. The Advertising Council alone in 1950 inspired 7 million lines of newspaper advertising stressing free enterprise, 400,000 car cards, 2,500,000,000 radio impressions . . . By all odds, it adds up to the most intensive 'sales' campaign in the history of industry." The campaign was used to "re-win the loyalty of the worker which now goes to the union and to halt creeping socialism, with its high tax structure and quasi-regulation of industry."[15]

This deluge of brainwashing paved the way to the shameful era of McCarthyism in 1950–54, when Senator Joseph McCarthy intimidated, hounded, discredited, and destroyed the lives of hundreds of his fellow US citizens, until finally one man, Joseph Welch, was brave enough to confront McCarthy in a Senate hearing and speak the truth. Soon thereafter, McCarthy died. According to Alex Carey, from 1950 to 1965 corporate power was again safe from the threat of a freethinking skeptical democracy. In 1955, *Fortune* magazine estimated that five thousand US companies had public relations departments, at an annual cost of about $400 million.

Do you see how the cold war was a logical spin-off from these successful initiatives to control domestic thinking? The somewhat contrived threats of Russia and communism, not innocuous by any means, of course, were used primarily to intimidate and silence the freethinking working people of the United States so that they lost their way and became confused when they should have been powerfully demanding their civil rights—free health care, education, and union power. The deadly nuclear arms race was conducted with impunity, protected by the camouflage that was created by this domestic manipulation.

Destruction of Price Control

The same tactics of combining public relations with corporate political ideology were used by big business to obtain uncontrolled price increases. The campaign I am about to describe was a prototype of techniques that continue to be used to "manage" democracy in the interests of American business.

After World War II, President Harry S. Truman was worried about rising prices because goods were in short supply. He decided to keep prices low by supporting and extending the life of the federal Office of Price Administration (OPA), which had been established during the war. But business wanted prices to rise. Represented once again by the NAM, it therefore launched a massive campaign against the OPA, printing millions of leaflets that were stuffed into the shopping bags of housewives. It also published full-page ads that stated cleverly but falsely that price controls themselves were the cause of the goods shortage. In 1946, the Opinion Research Council, monitoring the results, found that at the beginning of the propaganda campaign, 81 percent of the American people had favored the OPA, but that at the end, only 26 percent supported a continuation of the OPA. Americans had been manipulated yet again to act against their own best interests.

A discouraged Truman said, "Right after the end of the war, big business in this country set out to destroy the laws that were protecting the consumer against exploitation. This drive was spearheaded by the NAM."[16] In its effort to kill the OPA, the NAM spent $3 million, which in those days was a lot of money. As a direct result of this operation, consumer prices rose 15 percent and food prices 28 percent between June and December 1946. The people had again been exploited.

To add insult to injury, these price hikes canceled the wage increases that labor had obtained from some of its more successful 1946 strikes; at the same time, real wages dropped from $32.50

per week to $30 per week, while yearly corporate profits reached their highest point in history, $12 trillion—20 percent higher than those of the best war year.[17]

It is interesting that, since 1918, the Soviet government had brainwashed its people by consistently lying to them, but its techniques were so clumsy that the people knew they were being brainwashed. By contrast, in the United States, corporations had become expert manipulators, so most people swallowed the corporate doctrine whole.

Human Relations—Industrial Psychology

The corporations developed another nifty trick to convert their workers from "unionism" to "corporatism." It occurred to them that since most US workers were captive audiences in their factories, if they appealed to them the right way they could win their hearts and minds. Psychologists might be interested to know that the human relations movement was pioneered by corporate America for an ulterior motive. "Human relations"—a euphemistic phrase—was also called "employee participation," "employee communication," and "democratic decision making."

During 1945–46, business firms invested huge amounts of money in the study of this psychological discipline, and a plethora of books and literature appeared on the subject. Psychologists and social scientists were recruited to develop new and more effective methods to include workers in a science called "interpersonal communication," which was really used to induce workers to support their corporate bosses. By 1950, management had become obsessed with employee "communication." *Fortune* magazine noted, "There is hardly a business speech in which the word is not used."[18]

These techniques of worker manipulation proved a successful tool for bypassing union power in the factories, and worker loyalty swung toward management. In 1959, Peter Drucker, who represented American management consultants, said of human relations policies, "Most of us in management have instituted them as a means of busting the unions. That has been the main theme of these programs. They are based on the belief that if you have good employee relations, the union will wither on the vine."[19]

The literature on human relations continued to grow. During the decade of the fifties, there were four times as many studies of small human relations groups published in social science journals as in all previous publication history. Surprisingly, few sociological studies document the impact of this movement.

The Reagan Era

For fifteen years after Joseph McCarthy died, the American public was once again placid and under control. Most people worked hard for their friendly corporation—it was like one big family where loyalty reigned supreme. But then came the civil rights movement, Vietnam and flower power, Woodstock, and Watergate, and the nation once again lost respect for corporate control. In 1985, 20 percent of the 250 million people owned 44 percent of the money in the United States, creating vast disparities between the very rich on the one hand, and the middle class and the poor on the other.[20]

In 1975, the Advertising Council therefore launched another "economic education" campaign. Two years later, *Fortune* described the council's continuing campaign as "a study in gigantism."[21] By 1978, according to a congressional inquiry, US business was spending $1 billion a year of tax-deductible money on "education," to convince people that big government was bad for them. (The truth is that government regulation is bad for corporations. It is amazing how corporate advertising can turn truth on its head.) The campaign was once again successful, and public support for the idea that government is bad rose from 42 percent in 1975 to 60 percent in 1980. On the coattails of this expensive propaganda exercise, the doyen and figurehead of right-wing corporate America—Ronald Reagan—was elected. What an incredibly successful campaign! I remember it well. In 1975, the concept of big government was not a topic of discussion; by 1979, TV reporters would ask me, "But isn't big government bad?" I had no idea where this concept had come from. Now I know!

Under the guidance of George W. Bush, the philosophy and implementation of deregulation reached epic proportions. One of the significant reasons that the subprime mortgage debacle was able to fester in the first decade of the twenty-first century and almost destroy the US economy, along with the economy of much of the rest of the world, is that laws that regulated banks and financial institutions were by that time almost nonexistent. The Bush administration was one of the greatest proponents of deregulation and the privatization of government assets, including the sacrosanct social security system. The administration failed in the latter enterprise.

Here's a picture of the economic circumstances that prevailed before and during the Bush administration: In the 1990s, 1 percent of the wealthiest Americans held one-third of all the wealth in the US economy and earned 14 percent of the national income. Between the years 1979 and 2004, only 1 percent of Americans saw their income triple after taxes, 21 percent of those in the middle class saw their income increase by one-fifth, while the poor experienced no improvement.[22]

Significantly, in 2005, 35 percent of the members of Congress were millionaires. The average Senate campaign in 2006 cost about $5.8 million,[23] and the 2008 election campaign may well have exceeded $3 billion in television advertising alone.[24] Of course, given these figures, Congress largely excludes people who are not exceedingly wealthy, or who are not in the tier of society with influential and wealthy friends, making it difficult for the United States to have a truly representative democracy.

Treetops Propaganda and Think Tanks

Brainwashing entered a more sophisticated phase in the 1970s. Until then, the propaganda offensives had been "grassroots," but now the corporations decided to establish a series of "think tanks" staffed by brilliant, erudite people.

Although some private think tanks, such as the Conference Board and the Hoover Institution at Stanford University, have existed for several decades, some new, aggressive, right-wing tanks producing an incessant flow of market-oriented studies were established in the 1970s. Among them are the Heritage Foundation, the American Enterprise Institute for Public Policy Research (AEI), the Center for Strategic and International Studies at Georgetown University, and Business Roundtable. Funders include such reputable corporations as Reader's Digest, Hertz, Coors, Holiday Inn, Ocean Spray Cranberries, Bechtel, Gulf Oil, Vicks (makers of VapoRub), Amway, Hunt Oil, and the Chicago Tribune Company.[25] (Of course, there are also a number of think tanks that might be described as left-wing, among them the Brookings Institution, the Institute for Policy Studies, and the World Policy Institute, but they exert relatively little influence on the public agenda.)

These think tanks virtually created the new conservative movement of the 1970s and set Reagan's agenda. The Heritage Foundation drew up a comprehensive list of agenda items for Reagan's first and second terms of office. The first document was called "Mandate for Leadership—Policy Management in a Conservative Administration." During his eight years in office, Reagan withdrew financial support for the United Nations (since 1998, the United States has owed the United Nations over $1 billion in dues);[26] undermined the trade unions; mined the national parks; decreased funds for education, medicine, job training, community development, the poor, the elderly, and the indigent; stacked the US Supreme Court with conservatives; built more nuclear weapons to develop "superiority" over the Soviet Union; attempted to create the capacity to fight and "win" a nuclear war; gave tax breaks to the rich and increased the sales tax; and undermined the US Commission on Civil Rights. But the actual agenda of these corporate-funded think tanks is to (1) decrease government regulation of big business, (2) decrease taxes for corporations and for the rich, (3) destroy the unions, and (4) increase profits—all of which parrot the agenda of the Chicago economist Milton Friedman.[27]

In 1977, the AEI produced fifty-four studies on right-wing agenda items, twenty-two forums and conferences, fifteen analyses of important legislative proposals, seven journals and newsletters, and ready-made editorials sent to 105 newspapers. Public-affairs programs were carried by three hundred TV stations, and display units were produced for three hundred college libraries.[28]

Business Roundtable, founded in 1972, comprises 160 chief executive officers from the largest US corporations. In financial terms, the total revenues of these companies in 1981 equaled about half the GDP of the United States, and more than the GDP of any other country in the world.[29] As I write at the end of 2008, the total annual revenue of these companies is $4.5 trillion. In 1972, Justice Lewis Powell, a Nixon appointee to the US Supreme Court, urged business "to buy the top academic reputations in the country to add credibility to corporate studies and give business a stronger voice on the campuses." This happened. In the 1970s, business established chairs of "free enterprise," filled with hand-picked candidates, in forty colleges. What a prostitution of classical education![30]

Business Roundtable maintains a statesmanlike image, but according to Ralph Nader, "the dominant purpose leading to the foundation was a desire to combat and reduce union power," and "it proclaims moderation while sabotaging moderate reform.[31] Although Business Roundtable specializes in propaganda intended to manipulate legislation, it also works closely with the NAM and the US Chamber of Commerce in their grassroots activities. Together, they defeated labor law reform in 1978, which had been established to help reinforce the United States' declining labor unions, and they worked to oppose important consumer protection bills.

To this end, Business Roundtable hired a public relations firm that distributed canned editorials to one thousand daily papers and twenty-eight hundred weeklies, along with cartoons that attacked consumer protection bills. It also utilized a fraudulent poll claiming that 81 percent of all US citizens opposed consumer protection, when independent polls showed that one out of every two people favored the consumer protection bill that had been proposed in Congress at that time. This deceptive poll was published in a full-page advertisement in the *New York Times*. According to *Fortune* magazine, the defeat of this bill was a signal of victory;[32] in retrospect, it marked a watershed in the history of consumerism (its fate mirrors the defeat of Truman's Office of Price Administration in the 1940s).

So the business of America is business—as demonstrated in the successful campaigns of 1919–21, 1946–50, and 1976–80. To quote Alex Carey, "Complete business hegemony over American society was established. On each occasion similar, if not more sophisticated propaganda and public relations techniques were used." But what about the ethics of this serious situation?

Corporate America produces editorials, TV news pieces, and legislative material that is easy to understand, well conceived and written, and acceptable to both the media and Congress. The material is provided in a timely fashion to guide legislation on a particular issue. This sophisticated, high-level manipulation is called "treetops" propaganda. Instead of being directed toward the person in the street, it is focused on influential decision makers in Congress and in the media—newspaper editors, columnists, and television executives. Its immediate purpose is to set the terms of debate and to determine the questions and agenda that dominate public discussion.

Here are some of the new terms of debate:

- In the wealthiest country on Earth, should unemployment be maintained at 6 percent or at 10 percent? Not, should unemployment at any level be unacceptable? (Unemployment is good for business because it weakens unions' negotiating power by providing a pool of unemployed workers.)
- Should private doctors have more control over the medical system so that doctors make more money and only the rich get good treatment? Not, does every person have a right to free state-of-the-art treatment?
- Is it economically desirable to eliminate CFC gases? Should CFCs be reduced to 80 percent production by 2050, or would business lose too much money? Not, Should CFCs be eliminated completely?[33]
- Would auto companies suffer too much if they made fuel-efficient cars? Not, are fuel-efficient cars a necessity for saving the planet?

The Catholic Church has always said that if you capture people's minds and hearts in the cradle, you have them to the grave. It is the same with American business. People are born into this corporate atmosphere, and they die embracing it. This loss of individual freedom occurs in the greatest country on Earth, where "freedom" reigns supreme. I once asked a young woman why she thought the United States was the greatest country on Earth, and she said, "Because we're free." When I asked her, "Who runs the Congress?" she replied, "The corporations." So I pressed, "What do you really mean?" and she answered, "We have the freedom to get rich." I suppose this is partly true, because the nation is, or was, so extraordinarily wealthy. Money does trickle down from the very rich, so most people, before this 2008 economic meltdown, could afford one, two, or three cars, refrigerators, houses, and so forth. Even the poor usually have a refrigerator and a car. Yet, despite the overall appearance of affluence, most people in the United States cannot afford to be sick. Even those who are covered by private health insurance are often not adequately covered for a major illness; when they die, they leave their families destitute. This situation is discriminatory.

The conservative think tanks named earlier are involved in "policy research" or "agenda setting" for the corporate benefit. Their goal is not to save the Earth or to care for the American people but to enable the rich to get richer and maintain their power. I find it extraordinary that the rich expend so much effort and energy to gain ever more money and power, for these assets do not by themselves lead to happiness.

US Recession and Debt

What has happened to the United States as a result of this social engineering? The dynamics are complex, but in a certain sense the Reagan years saw the culmination of the unbridled corporate activities that became the basis of the Bush years.

The truth is that the United States is not really a supremely rich country any longer. Its foreign debt is over $6 trillion, 23 percent larger[34] than it has ever been, and if Japan and China were not now economically prepared to support the US foreign debt, American society would collapse, along with much of the rest of the Western world. The US national debt today is almost $12 trillion, and it is growing at the rate of $1 million a minute. In 2002, just a year after taking office, George W. Bush borrowed $133 billion, whereas Clinton had borrowed only $18 billion in his last year in office. President Bush set numerous records for increasing the nation's debt. In 2003 alone, he borrowed over $500 billion. Even President Reagan never increased the debt so significantly in one year. In 2006, President Bush announced that he was borrowing $300 billion a year, but records show that the figure was twice that.[35] One of the main historical reasons that led to the weakening of the US economy is that the Reagan administration spent more money than all prior US administrations combined. And most of it went to build weapons of mass destruction. In fact, "Reaganomics" led the United States into a deep recession, converting it from being the biggest creditor nation to being the biggest debtor nation in history at that time. George W. Bush, however, surpassed the Reagan era in spending.

The recession during the Reagan era and beyond was aggravated by the savings and loan (S&L) scandal, which was instigated when the Reagan administration deregulated the S&L industry. Crooked dealers moved in and made a fortune by investing government-guaranteed money in junk bonds and dicey real-estate deals. Many of the modern high-rise buildings in the Sun Belt—in Dallas, Houston, and other cities—were built with this money.

It was estimated that the S&L fiasco would cost the US taxpayers almost $1 trillion over the years following.[36] One of its prime instigators was Charles H. Keating, who ran Lincoln Savings and Loan. Keating's professed philosophy was this: "Always remember the weak, meek and ignorant are always good targets." Keating allegedly paid $1.4 million to five US senators, including John Glenn and Alan Cranston, in bribes on behalf of Lincoln Savings and Loan. These men, dubbed the Keating Five, were subsequently acquitted in a court of law. Keating was charged with criminal fraud for the $2.5 billion failure of Lincoln Savings and

Loan, and he faced a $1.1 billion civil fraud suit filed by the US government—the biggest suit in history.

Michael Milken was jailed for similar activities. The company Drexel Burnham Lambert, which financed junk bonds, was charged with a $6.8 billion lawsuit. Large numbers of ordinary people lost an enormous amount of money collectively—in many cases, their life savings.[37]

You see, the Reagan era introduced a rampant "me now" generation of corporate cowboys who made millions or billions of dollars by defrauding other people. US banks are currently in serious financial straits, partly because of the Third World debt and partly because of unwise and unregulated lending to these corporate cowboys, who often did not provide the banks with collateral. When property prices collapsed during the recession, banks discovered record bad loans on their books. Because the US government guarantees the deposits of failed banks, as well as of the S&L industry, it faced a crisis in the banking industry that threatened to be as massive as the S&L fiasco. For example, one of the largest banks in the United States, the Bank of New England, collapsed early in 1991, costing the US government $2.5 billion.[38] Indeed, a similar situation arose late in 2008, forcing the US federal government to utilize the government fund that guarantees the banks.[39]

Meanwhile, to add fuel to the fire in 1991, the US insurance industry was in deep trouble. Insolvencies were on the rise. The assets of the largest insurance companies were equivalent to the assets of the largest banks. In 1990, forty-four insurance companies failed, compared with thirty-one in 1989. The failures are attributed to unwise insurance company investments in real estate, junk bonds, and other areas. (The federal government does not regulate this industry, which in 1991 controlled $32 billion in real estate, $267 billion in real-estate mortgages, and $65 billion in high-risk junk bonds.)[40] Now, in 2008, the insurance companies are again experiencing difficult times because of the increasing frequency of natural disasters secondary to global warming; their premiums rose 46 percent from 2001 to 2006. The insurance premiums in Florida rose 77 percent; in Maryland, 76 percent; in Virginia, 67 percent; in Minnesota, 66 percent; and in Louisiana, 65 percent. Hurricane

Andrew in 1992 cost $26.5 billion in damages, and twelve insurance companies were forced to shut down. In 2004, Hurricanes Charley, Frances, Ivan, and Jeanne caused $40 billion in damages and $22.5 billion in insurance losses. The 2005 hurricanes, including Katrina, Rita, and Wilma, cost $55 billion in insured losses.[41] The insurance industry in 2008 now controls an estimated $335 billion in real estate, making it vulnerable to the subprime crisis as real-estate values crumble.

The eight years of Reaganomics encouraged a kind of frenzy among American moneymakers. People played the stock market, buying and selling junk bonds. They invested in futures markets, indulged in illegal insider trading, pursued acquisitions and mergers on a grand scale, set up cross-directorships, and gambled with shareholders' money. They also paid huge salaries to themselves and corporate executives. Many of these activities were illegal. These people seriously jeopardized the banks that lent them money, threatened the US insurance industry, and destroyed the S&L industry.

Amazingly, the United States now faces another similar catastrophe. The free reign given to the corporate cowboys during the eight years of the George W. Bush administration led to the subprime crisis. The *Wall Street Journal* predicted that by the end of 2008, roughly $600 billion in adjustable-rate subprime loans were due to adjust to higher rates—meaning that an increasing number of borrowers would fall behind in their loans, specifically mortgage payments, induced by irresponsible lending by banks and finance companies that were unregulated. In other words, the banks lent to unqualified borrowers, they inflated appraisals, and did not properly verify borrowers' incomes to support their loans. And as this financial catastrophe unfolds, the values of peoples' houses are dropping drastically, so hundreds of thousands not only cannot pay the ever-rising interest and principal on their loans, but they are going bankrupt.[42]

It is generally acknowledged that the subprime crisis in the United States has precipitated the worst global economic crisis since the 1930s. The two largest financial institutions in the world—the mortgage insurers Fannie Mae and Freddie Mac, which together owe $5 trillion—teeter on the brink of bankruptcy and, as in the Great Depression of the 1930s, people are lining up in queues outside banks to retrieve their money.[43] The federal government was impelled in September 2008 to bail out Fannie Mae and Freddie Mac, to the tune of $200 billion, thereby taking charge of the $5 trillion in home loans that these corporations had backed.

It really has been a case of the totally unregulated corporate cowboys gambling with impunity with other people's money. Let me give you an example, and I quote from an article in the *Times* of London by Anatole Kaletsky: "In the old world before the arrival of 'hyper-finance,' if a family wanted a $100,000 mortgage, they would simply go to the Halifax and borrow $100,000. Now consider what happens in the new financial world. The family would borrow $100,000 from Northern Rock, which would sell $100,000 of bonds to hedge funds, which would buy these with $100,000 borrowed from Bear Sterns, their prime broker, which would raise this money by selling $100,000 of commercial paper to Citibank, which would then borrow $100,000 through the inter-bank market from Halifax. So now the original $100,000 mortgage transaction has created $500,000 of new debts."[44]

The year 2006 was huge for global mergers and acquisitions, which totaled almost $3.5 trillion that year, compared with $3.3 trillion in 2000. AT&T's $83 billion buyout of BellSouth was the single largest transaction in 2006. Four out of five of the other largest transactions occurred in Europe. As a result, European stocks have become more liquid over the years.[45]

The US federal deficit is forecast to be $500 billion next year, and the Iraq war is costing more than $200 million a day.[46] When George W. Bush took office in January 2001, he inherited a US surplus of $5.7 trillion. By December 2007, that surplus had been transformed into a deficit of more than $9.13 trillion—about $30,000 for every American, or $1 million per minute.[47] Yet while the American economy is under siege, the Coca-Cola Company announced a 13 percent earnings increase in the third quarter of 2007, making a profit of $1.65 billion in those three months.[48]

Fancy making this sort of money from the sale of colored candy water in plastic bottles!

Where are our values, and what is the world coming to? Many multi- or transnationals continue to do well, despite the federal government's struggle to repair the economic shambles left as a legacy by corporations that took advantage of the lax federal regulations introduced by the Reagan and George W. Bush administrations. This situation is a perfect example of the end result of right-wing think tank policies. The Reagan and Bush presidencies marked a bonanza for big business and a tragedy for the United States, and who pays? The middle class and the poor—80 percent of the US population.

Pathology Induced by the Corporate Ethic

Insidious corporate proselytizing has inculcated into American workers an ethic that says, "Ask not what I can do for my country; ask what I can do for my corporation." In a certain subliminal way, the sense of country and the sense of corporation have been turned into one and the same. If people have been convinced that these corporate values are the norm, they do not expect that they, as individuals, have an inalienable right to live in a society of compassionate values where free health care, free higher education, and a fair tax system are universal. But people in countries such as Australia, New Zealand, Britain, France, Canada, and the Scandinavian nations experience these benefits and indeed take them for granted.

It is also becoming quite clear that most people in the United States realize that something is very wrong when people die in hospital parking lots because they have no health insurance, or when families become almost destitute trying to provide a college education for their children. It is also obvious that while the rich and corporations pay a relatively small percentage of their income in tax, the middle class and the poor bear the financial burden of running the country.

In fact, during the Reagan era, which brought the culmination of years of corporate planning, the rich gained and the poor lost ground because Reagan decreed that tax cuts for the wealthy would stimulate investment and expand the economy and that the middle class and poor would benefit when the profits eventually "trickled down" to them. Of course, a similar but almost more severe dynamic occurred under George W. Bush.

The trickle-down theory never worked. According to the Congressional Budget Office, the top 5 percent of all Americans received 45 percent more income before taxes in 1990 than they did in 1980, but during this time their tax rate decreased by 10 percent, from 29.5 to 26.7 percent. In contrast, the poorest 10 percent of the American people earned 9 percent less income in 1990 than in 1980, and their tax rate increased by 28 percent.[49] During the Bush administration, tax rates for middle-income earners inched up, while the very rich 1 percent benefited enormously as they received large tax cuts in investment income and a steady reduction in real-estate taxes.[50] Note also that most corporations do not pay income taxes, according to a Government Accountability Office report released in August 2008.[51]

A way to tax the poor indirectly, without actually saying so, is to tax everyday items like clothes, gasoline, cigarettes, and beer. I am always shocked when I buy a dress in the United States for $50 and the salesperson at the counter says, "No, it's $55 with tax." This practice can be defined as regressive taxation. In order to end this deceptive practice, the price indicated on the label should include the tax. Of course, the poor and the middle class are most affected by sales tax; the rich hardly notice it, simply because such expenditures make up only a tiny percentage of a wealthy person's spending.

In 1990, Congress was deeply concerned about the enormous deficit and knew that it must increase taxes, despite George Bush's "read my lips" speech. But members of Congress are always reluctant to tax the wealthy, and big business is in a good position because it funds congressional elections and lobbies Congress between campaigns by means of right-wing think tank strategies and staff.

I pay 40 percent of my income in Australian federal tax, but I do not object and am even quite proud, because I know this money will be used to care for and educate me and my fellow citizens. I am aware, though, that many people in the United States dislike paying taxes because they know that this money will generally not be used for their own benefit and well-being, much of it going to the Pentagon.

The IRS took in over $2 trillion in fiscal year 2007. Nearly 40 percent of that total went to military-related expenses, according to a report by the National Priorities Project. The research group estimated that 27 percent of individual federal taxes would pay for current military spending, including the war in Iraq. An additional 9 percent would help pay off debt from past wars and military expenses. And another 3 percent would cover benefits for veterans.[52]

Some 47 million Americans have no medical insurance.[53] More and more middle-class families can no longer afford the premiums, and the federal government does not provide free medical care. Ironically, despite these glaring deficiencies, the United States spends more on health per capita than does any other nation. According to a 2007 Commonwealth Fund report, the US system ranked below the Australian, New Zealand, German, Canadian, and British systems—each of which provides excellent health care for all its citizens. And a WTO report ranked the United States thirty-seventh out of 190 nations in health care services.[54] If these countries can do it, so can the United States, which has always been innovative and creative as a nation.

Let me give a classic example of the attitude of corporate business toward the health care of its workers. Industry leaders have become increasingly concerned about mandatory employer-funded health care schemes, saying that they are unfair to the private sector and contribute to higher prices—everything is passed on to the consumer while business always profits.[55] Of course, the government should be paying for universal health care instead and not wasting over $625 billion on weapons and the military-industrial complex.

Private charities—George H. W. Bush's "thousand points of light"—are meant to care for poorer people when government abrogates its responsibility. Soup kitchens, churches, and good people are expected to take up the slack. For years, Corporate America has encouraged the government not to provide social services for people. Instead, large numbers of tax-deductible charitable foundations have developed. They enable rich corporations to evade taxes by channeling money into charity and "educational" pursuits. Some of these foundations do actually care for needy people, but many people fall between the cracks. It is tragic that charity is expected to care for America's poor, sick, and indigent, while tax dollars are pumped into the Pentagon, especially now that the Soviet Union is no longer intact and the Communist Party has been outlawed by the Russian Federation. Indeed, President George W. Bush fostered and funded "faith-based" organizations, meaning that if you believe in God (usually the Christian God), you are more likely to be able to receive federal funds to look after the poor and sick. Surely this connection violates the separation between church and state!

In 1990, Congress tried to increase taxes for Medicare, which brought howls of rage from the elderly, along with other taxpayers. From 1990 to 2003, the inflation rate for health care costs rose by 41 percent, and overall health care costs increased by 82 percent. Medicare payments by the public increased by more than 300 percent from 1980 to 1990—six times the rate of inflation—and they continue to rise. In September 2007, Congress passed a bill that temporarily postponed deep cuts in Medicare payments by patients, increasing Medicare physician payments for the first six months of 2008 by 0.5 percent, thereby postponing a 10.1 percent reduction in payment rate.[56]

The medical profession is a very powerful lobby, not for the patients' benefit but for its own. Remember that Medicare provides little money for prescription drugs or for protection against financial ruin from the cost of nursing-home care. Many old people are therefore forced to deplete their life savings in order to pay for medical care. This burden is particularly tough for the elderly poor—the 23 percent of those over sixty-five who live on an annual budget of $10,000 or less. The median income of people sixty-five or older in 2006 was $16,890.[57] In 1999, per capita health care spending for the average American was $3,834. For those under sixty-five, it was $2,793; but for those over sixty-five, it was $11,089—almost four times higher.[58] This number includes nearly half the older women in the United States, and two-thirds of the elderly minorities. As residents of the richest country in the world, these people deserve a measure of comfort and dignity in their old age.[59]

Australia has a dual medical system: free medical care for all, and a voluntary private health insurance scheme that enables individuals to choose their own doctor and private hospital. Acutely sick people are actually much better off in a public hospital with resident medical staff and excellent twenty-four hour laboratory service. Private hospitals have nice rooms and nice food, but private-hospital patients have been known to bleed to death at night without around-the-clock medical cover. In other words, Australian tax dollars are used for world-class medical care. I had my gallbladder removed by the best surgeon in town. I spent ten days in the hospital and enjoyed excellent nursing care and tasty meals, and I paid not one cent. It was all on the government, funded by citizens' taxes.

The US medical system, by contrast, is almost totally privatized. I make these comparisons not to put the United States down, but to suggest ways to foster a more compassionate society. The United States has created a large and advanced medical system capable of treating every known illness, yet people die in their homes because many cannot afford to see a doctor and be admitted to a hospital—a standard hospital bed can cost $8,500 for twenty-four hours, when the average minimum American wage is only $7.25 per hour. This health care system concerns itself primarily with treating the rich. The United States is virtually the only industrialized country that fails to offer comprehensive medical care to all its citizens.[60]

Many hospitals are run by private corporations and entrepreneurs who see medical care as a profit-making venture. Nursing homes have become a huge "industry"—fancy calling the medical care of old people an industry, or do we make money out of just anything? The owners open and close hospitals at whim, depending on whether they are making money or not, hiring or firing nursing staff, admitting or discharging patients. Medicine should not be practiced for profit; it is a service that should be offered with humility and compassion.

I remember when the president of the American Medical Association said in 1969 that medical care was a privilege and not a right. This attitude violates the Hippocratic oath, a code of compassionate medical ethics—the very philosophy of medical practice. My first job as a newly qualified doctor was to relieve a country physician for two weeks while he went on vacation. He paid me several hundred dollars for my time, and I felt embarrassed to receive money for the privilege of practicing medicine. Medical care is a responsibility for doctors and the community. All people—clean or dirty, young or old, rich or poor—have a right to the best medicine that society can offer.

Education

Finally, I want to address the important subject of education, because to a certain degree the fate of the Earth rests upon the education that children are now receiving.

In Australia, all universities, until recently, were funded by the federal government, and education for every discipline, including law, medicine, architecture, and science, was totally free. But some twenty years ago, the government introduced a small tertiary tax of $1,000–$2,000 per year, which marked the beginning of the privatization of our university system, and the fees for a university education have progressively risen since then. This new scheme was and is strongly supported and encouraged by right-wing think tanks employing the US treetops philosophy and by corporate Australia. But my daughter, who graduated from medical school in the late 1980s, had almost her entire education paid for by the Australian government. My medical education was also totally covered by the federal government. What a gift! In Australia, high schools and primary schools are funded and run by the state governments, and education is free and uniform throughout each state. Like health care, free education is deemed a right of all children and people.

When I lived in Boston, I was surprised to discover that school districts in the United States are funded largely by local government with federal and state support. It follows that if schools are located in poor communities, the educational standards or facilities will be correspondingly low. In the United States, private schools are very expensive and available mainly to the middle class and the wealthy, as are many good colleges and universities—with the exception of some excellent state systems, like that of California. It is hard to believe that it can cost $40,000 or more to send one child to college for one year. For a family of six, educating four children breaks the budget and often leaves the parents in penury for years. Such a state of affairs is not fair, equitable, or right. It is also imperative that we raise a generation of adults who have been well informed about the parlous state of the ecosphere from both a scientific and a political perspective. Only through education will these people develop the wisdom and ability necessary to save the planet.

Another problem worries me. I speak regularly at hundreds of colleges across the United States, and each year I give several commencement addresses. I find that most college presidents are primarily fund-raisers who have little time to devote to the curriculum or to education per se. Fund-raising is not an easy task; I know because I used to raise $3 million a year for my two organizations—Physicians for Social Responsibility and Women's Action for Nuclear Disarmament—from philanthropies. Colleges receive money from wealthy individuals and from foundations and corporate bodies, and their curricula are often influenced by the wishes and philosophies of the funders. This is not free Socratic education. Often when I'm standing at the entrance of a college, I see plaques acknowledging generous gifts from IBM, GE, and other corporations, and I worry.

Apart from some of those at exclusive Ivy League schools, I find most American college students to be almost totally ignorant about world geography, world history, English literature, the true history of the United States, or the history of the nuclear age. Most do not know about Hiroshima and Nagasaki or about Hitler, and they do not have an in-depth understanding of the etiology of environmental threats to the Earth.

A 2006 *National Geographic* survey of 510 students between eighteen and twenty-four years of age revealed that six months after Hurricane Katrina, 33 percent could not identify Louisiana on the map, 88 percent could not locate Afghanistan, 63 percent could not find Iraq or Saudi Arabia, 75 percent could not find Israel or Iran, and 42 percent were unable to find any of the countries listed. In fact, half or fewer young men and women 18–24 can identify the states of New York or Ohio on a map of the US.[61] The United States ranks forty-ninth in the world in literacy, and 23 percent of Americans think the sun orbits the Earth.[62]

Many of the American students I encounter have received little or no education in the ethics of business, and often they do not know what job they will get, except that it will be in the corporate world. I ask women who are graduating from business school, "What will you do with your degree?" and they say, "Oh, I want to go into marketing." When I ask further, "What will you market?" they say, "Oh, anything." And that means really anything!—disposable diapers, CFC-propelled spray cans, cruise missiles—you name it. The engineering and mathematics graduates can find virtually no job except in the military-industrial complex, which has recently received a big boost from the Iraq war. (However, during the current global economic crisis, many students are shunning business studies and are attracted more to education and public affairs and administration.)[63]

These poor kids have not been given a comprehensive understanding of the world in which they live. If they are ignorant of the dynamics of the world, how can they protect their future? Furthermore, I find that they do not know how to think critically. When I give a lecture and tell them facts that are totally alien to anything they have ever heard before, they ask, "But how can we believe you?" and I say, "Don't believe me; do your own research and find the truth for yourselves." They seem perplexed and bewildered at the enormity of the task. They have not been taught how to use a library. I sense that most of them have been nicely conditioned and programmed to be conveniently pumped into the system of corporate America. Remember, it is the corporations in the Western world that are mainly responsible for global pollution, species extinction, and the threat of nuclear war. I am not excluding the dreadful mess the communist regimes have made of the environment in eastern Europe, China, and the ex–Soviet Union.

Many college exams are multiple-choice and require little use of the English language. College teachers often despair as they tell me that students are poor spellers and have woeful math and reading skills. The teachers blame the primary and secondary educational system.

I think it would be better if students were forced—and I mean *forced*—to study a particular subject in depth, so that they learned to use the five layers of neurons in their cerebral cortex. They really need to understand the concepts of critical thinking and research, and honest questioning—in order to develop a degree of skepticism toward what they watch on TV and read in the papers. Unfortunately, though, most articles in newspapers now resemble TV news, offering small "read bites" instead of "sound bites." I also find that most students do not read newspapers. Indeed, they tend to spend all their time on cell phones texting each other and on computers, which provide a hodgepodge of often superficial information that is very difficult to navigate. It is almost impossible for students to gain an in-depth analysis of any subject using these new—and, I think, rather dangerous—technologies in terms of understanding the world that they are about to inherit.

The standard of education should be uniformly high for all students, rich and poor, throughout the land. The federal and state governments must pay well to educate the next generation, and teachers must be free and independent of corporate and big financial interests. Some from this uneducated student body will emerge as the teachers of the next generation, thereby compounding the problem.

The ideal educational system in many ways parallels the ideal health care system; both should be free and paid for by society's tax dollars. However, education is more vital for a nation. Our children are inheriting a dangerous world; dying from pollution, global warming, and overpopulation; wired up by transnational corporations like a ticking time bomb ready to blow up at any minute from a nuclear war. Yet they really have little knowledge of the situation, except for a vague understanding they obtain from watching TV news.

It is absolutely imperative that they be well informed in order to ensure their own survival. Teachers, I believe, are the most responsible and important members of society because their professional efforts will affect the fate of the Earth. Therefore, they should be paid a high salary so that the best people become teachers.

10 The Media and the Fate of the Earth

Let's take a searching look at the media, which are, we may safely say, determining the fate of the Earth. If newspapers and TV in past years had decided to publicize ozone destruction, global heating, deforestation, species extinction, overpopulation, and the threat of nuclear war as they publicize AIDS, they could have educated, inspired, coerced, and persuaded people to save the planet. Too often, though, we received only superficial or sensational news stories, such as those involving rape, murder, fires, and violence, which sell papers and advertising on TV. Corporations that own the media, on the whole, are out to make money and to control the public through a perpetuation of ignorance, not to educate. As the right-wing talk show host Rush Limbaugh said, "First and foremost I'm a businessman. My first goal is to attract the largest possible audience so that I can charge confiscatory ad rates. I happen to have great entertainment skills, but that enables me to sell airtime."[1]

At journalism schools, however, reporters are taught the ethics of journalism—fairness in reporting and thoroughness in investigative research. Often, good reporters and their articles are quashed by editors representing megamedia bosses like Rupert Murdoch.

In 1982, fifty corporations owned just over half the media business in the United States; by 1987, the count was twenty-six; and by 1990, it had decreased to twenty-three. In 2000, AOL and Time Warner signed a $350 billion merger. *Mother Jones* reported in 2006 that only eight giant media companies controlled most of the US media. Today, these conglomerates include AOL–Time Warner, Disney (ABC, Disney Channel, ESPN, 10 TV stations, and 72 radio stations), Bertelsmann, Viacom (CBS, MTV, Nickelodeon, Paramount Pictures, Simon & Schuster, and 183 US radio stations), News Corporation (FOX, HarperCollins, *New York Post, Weekly Standard, Wall Street Journal, TV Guide*, DIRECTV, and 35 TV stations), TCL, General Electric (owner of NBC), Sony (owner of Columbia and TriStar Pictures), and Seagram (owner of Universal Studios).

Most online news services today do not produce their own information, but instead reprint it from Reuters and the Associated Press, even mixing some of the content together from such agencies.[2] As Phil Donahue, who was fired from MSNBC in 2003 for interviewing antiwar people, said, "We have more TV outlets now, but most of them sell the Bowflex machine. The rest of them are Jesus and jewelry. There really isn't any diversity in the media anymore. Dissent? Forget about it."[3] One billion people around the world tune in to CNN International every day, and it is broadcast in 212 countries.[4] To put it simply, many of the important transnationals are interrelated and will therefore work cooperatively. Such a situation is terribly dangerous. Economic control, political control, and mind control—these are just extensions of the GATT philosophy transferred from the Third World to the First.[5]

Apart from securing political and economic control, the primary motive of most media companies today is to make money. They do this by selling advertising, and if the corporate sponsors who buy the ads do not like the program content, they cancel the ads. So the program content is, in effect, largely controlled by the advertisers.

We must also examine the corporations that now own the US media and their connections with other business. For instance, General Electric, which manufactures nuclear power plants, nuclear weapons, and missiles, also owns the National Broadcasting Corporation. It is surely fair to ask, therefore, whether NBC could be impartial in its analysis and reporting on nuclear power stations, radiation accidents, demonstrations against nuclear weapons testing, détente, or the Iraq war. Impartiality is nigh impossible.

General Electric may serve as a prototype transnational corporation that has an enormous impact through the media. You might think that GE is true to its motto and brings "good things to life," like irons, stoves, washing machines, and refrigerators (all of which use electricity). But what is this corporation really doing behind its benign facade? Its operations extend into fifty countries in its search for markets, production facilities, and raw materials. Ronald Reagan was GE's devoted salesman for some ten years, as the host of the *General Electric Theater* television programs from 1954 to 1962. GE built an electric house for the Reagans in the 1960s, complete with such new inventions as a garbage disposal unit and a dishwasher.[6]

As stated already, GE has been involved in nuclear weapons production since the end of the Second World War, as well as in the construction of nuclear power plants. In 1945, GE's president, Charles Wilson, opposed conversion of the military economy to civilian production and helped set in motion the machinery to ensure a permanent war economy. Because of his early actions, GE had by 1991 become one of the largest nuclear weapons producers in the land, grossing $11 billion in nuclear warfare systems in the period 1984–86. GE has made parts for the Trident and MX missiles and for the Stealth and B-1 bombers. It has been the developer and sole producer of the trigger for every nuclear weapon made in the United States; it manufactures Strategic Defense Initiative (SDI), or "Star Wars," components, and it has had a key role in the manufacture of all nuclear weapons (each one costs $40 million, and at the height of the cold war, five new bombs were manufactured every day), ranging from uranium mining to plutonium production, weapons testing, and nuclear waste storage.

In 2004, GE purchased Vivendi's television and movie assets, named NBC Universal, making it the largest media consortium in the world. In 2005, GE established its $90 million "Ecomagination" campaign to market itself as a "green company," by producing hybrid locomotives, desalinization and water-recycling solutions, and photovoltaic cells. In only a year, revenues exceeded $10 billion. Within the coming year, GE plans to sell its technologies to companies that are not easily adjusting to stricter environmental regulations.[7] Large corporations are now realizing that they must hop aboard the environmental train because not only will they be doing the right thing, but they will also make a tidy profit.

GE executives also belong to key business groups and think tanks that exert enormous influence on government policy, including the Business Council, Business Roundtable, and the Council on Foreign Relations. In addition, they are members of exclusive social clubs, where they fraternize with the elite and powerful—the Bohemian Club in California, the Economic Club and the Links Club in New York, and the Augusta National Golf Club in Georgia. It is often within the confines of these clubs that some of the most important political decisions are made. Not least, GE executives belong to very influential Pentagon committees. In the past, for instance, one executive who held various positions in GE headed a presidential space commission in 1987 that strongly recommended that NASA develop a space station, and in that same year GE was awarded an $800 million contract for work on the station. The chairman of GE, John Welch, was, until 1990, also chairman of the National Academy of Engineering. GE aggressively lobbies for its weapons systems. In fact, the company has had more registered lobbyists than has any other weapons contractor.

Another network, ABC, is owned by Capital Cities, a huge company with interests in radio and publishing. But who owns Capital Cities? William Casey, the now deceased director of the CIA under Reagan, founded Capital Cities in 1954. When asked to put his stocks in a blind trust in 1983 because of his administrative appointment, Casey quietly kept control of his single largest stock: $7.5 million in Capital Cities.

The history of the purchase of ABC is very interesting. In November 1984, Casey, as CIA director, asked the Federal Communications Commission (FCC) to revoke all of ABC's radio and TV licenses because one of ABC's news reports suggested that the

CIA had attempted to assassinate a US citizen. In February 1985, the CIA asked the FCC to apply the fairness doctrine to ABC; in March, Capital Cities Communications bought ABC. For the newly acquired ABC, this was not a good beginning with regard to impartiality and fairness in reporting. It was now owned, in effect, by the head of the CIA. Other board members of Capital Cities at that time sat on the boards of, or were connected with, IBM, General Foods, Johnson & Johnson, Texaco, Avon, Conrail, and many others. Do you see the interconnecting links between transnational corporations and the media?[8]

Without a free and uncensored press, there can be no legitimate democracy. If the people do not have the relevant facts, they cannot make informed decisions about their politicians, their country, or their world. Secrecy is promoted to maintain the power structure of the occasional woman and the white males who run the world. Many major decisions that affect our future are made behind closed doors, and then we are manipulated by the media to believe what we are told. The 1990 war in the Persian Gulf and the present Iraq war offer good examples of this strategy.

If society demands full financial disclosure from its politicians, it must demand the same from the corporations that own the media. Proposed legislation that would require disclosure of financial assets and ties would be particularly pertinent when documentaries and stories are produced that could affect the profits of the corporate parent or the sponsors. These shows often do not go on the air or are not published. Furthermore, the actual ownership of a company must be made public. We tend to believe that we are dealing with a respected and well-known company, but because of corporate takeovers, the organization may really be owned by a bigger, silent corporation, which probably cares little about the quality of the product but very much about profits.

In fact, only 118 people in June 2005 made up the membership of the ten big media giants in the United States. These people, in turn, sat on corporate boards of 288 national and international corporations.

Here are but some of the corporate board interlocking networks for the US media giants:

- *New York Times*: Carlyle Group, Eli Lilly, Ford, Johnson & Johnson, Hallmark, Lehman Brothers (until recently), Staples, and PepsiCo
- *Washington Post*: Lockheed Martin (the biggest weapons manufacturer in the world), Coca-Cola, Dun & Bradstreet, Gillette, GE Investments, J. P. Morgan, Moody's
- Knight Ridder: Adobe Systems, Echelon, H&R Block, Kimberly-Clark, Starwood Hotels
- Tribune (*Chicago Tribune* and *Los Angeles Times*): 3M, Allstate, Caterpillar, ConocoPhillips, Kraft, McDonald's, PepsiCo, Quaker Oats, Schering-Plough, Wells Fargo
- News Corporation (FOX): British Airways, Rothschild Investments
- GE (NBC): Anheuser-Busch, Avon, Bechtel, Chevron/Texaco, Coca-Cola, Dell, GM, Home Depot, Kellogg, J. P. Morgan, Microsoft, Motorola, Procter & Gamble
- Disney (ABC): Boeing, Northwest Airlines, Clorox, Estée Lauder, FedEx, Gillette, Halliburton (before becoming vice president under George W. Bush, Dick Cheney was the CEO), Kmart, McKesson, Staples, Yahoo
- Viacom (CBS): American Express, Consolidated Edison, Oracle, Lafarge North America, Gannett, Associated Press, Lockheed Martin, Continental Airlines, Goldman Sachs, Prudential, Target, PepsiCo
- AOL–Time Warner (CNN): Citigroup, Estée Lauder, Colgate-Palmolive, Hilton[9]

Then there is the Public Broadcasting System, or PBS, which many people consider to be an impartial public network. But over the years it has become partly privatized by default. Many of the fine programs on PBS are sponsored by corporate polluters that are in trouble because they have added to the toxic woes of the world and want to redeem their image. Because of these sponsors, PBS is facetiously referred to now as the "Petroleum Broadcasting Service."[10]

Both PBS's redoubtable *NewsHour with Jim Lehrer* (formerly *The MacNeil/Lehrer NewsHour*) and ABC's *Nightline*, anchored by Ted Koppel until 2005, were and are seen as models of excellent, responsible investigative journalism. But Fairness & Accuracy In Reporting (FAIR), a media watch group, analyzed the two programs' guest lists and found some rather disturbing trends. According to FAIR's six-month study, released in April 1990, both programs interviewed a disproportionate number of white males. On the *NewsHour*, 90 percent of interviewees were white and 87 percent were male. For *Nightline*, the comparable figures were 89 percent and 81 percent. The percentages were even higher when international politics was discussed. *Nightline* featured environmental issues in only 6 of 130 programs. Of the fifteen guests on these environmental shows, all were white, and only two were women—one of them Margaret Thatcher. Nine of the fifteen were government officials, and two were corporate representatives. Only two were environmentalists.[11]

Ralph Nader said recently, "Look at all the stories on the destruction of the Amazon forest. Do you ever see the names of any multinational corporations mentioned?" Forest destruction and other environmental disasters all seem to happen spontaneously, and no one is held responsible on these shows.[12]

When Robert MacNeil was asked some years ago by FAIR why his program tilted toward the right, he replied, "There is no left in this country"; and Jim Lehrer responded to suggestions that appropriate critics of government policies be given airtime by calling them moaners and whiners. On seven *MacNeil/Lehrer NewsHour* segments on the *Exxon Valdez* spill broadcast in 1989, not one environmentalist appeared. Ted Koppel once said, "Policy critics aren't needed on *Nightline* since we invite the policy makers and ask them the 'tough questions.'"[13]

On these programs, women (who make up half the population), people of color, and union members are not represented. Once again, a very small, powerful, unrepresentative section of the community dominates the airwaves, the debate, and the agenda.[14]

Hawks are put up to debate hawks, the right to debate the far right. For instance, Caspar Weinberger, Reagan's secretary of defense, was pitted against Senator Sam Nunn, a hawkish senator in those days, on the nuclear arms race. Why not Randall Forsberg (who invented the nuclear freeze concept and was one of the leaders of the peace movement of the 1970s and '80s) and Weinberger, or Forsberg and Nunn? At least it would have been a lively debate and, above all, an informative one.[15]

Just before the 2003 Iraqi invasion I was interviewed for half an hour by a camera crew from *The Oprah Winfrey Show* about the state of the world. During the interview, which was conducted very carefully, I said that "we have never been closer to nuclear war." How then did they use my interview? Only several seconds of this statement was played; I was portrayed as a gray, grainy image; and they pitted me against two hawkish men in the studio who said yes, that's why we need to go into Iraq! This is an example of the media propaganda that led to the American people's support of the shock-and-awe invasion.

Meanwhile, PBS has become more and more business-oriented over the years, as it has fallen into the corporate orbit. It now runs the *Nightly Business Report*; and until recently, its programming included *Adam Smith's Money World* and *Wall $treet Week*. These programs were sponsored by Metropolitan Life and Prudential Bache. PBS obviously receives money by hosting these entirely pro-business shows and it has, to a certain extent become a handmaiden to business, thereby losing its editorial independence. For the last several years both PBS and NPR (National Public Radio) have carried ads for the Nuclear Energy Institute advertising the wonders of nuclear power—clean, safe, with no emissions. (These statements are pure fabrications and very dangerous, so I called the head of NPR to complain, but to no avail.) Fancy the national broadcaster prostituting its soul for advertising money, as this industry will kill untold numbers over future generations from its radioactive waste. Just like any other broadcasting network, PBS is, to a large degree, controlled by and answerable to its corporate sponsors.[16]

Corporations are rarely exposed, and workers get a sympathetic press only in the context of environmental issues. If environmentalists try to save forests, the workers' jobs become paramount. Typical headlines are "Old-Growth Forests versus Loggers," "Trees versus Jobs," and "Save the Planet versus Save Our Jobs."[17] The situation is similar in Australia. The forests must be destroyed to provide jobs for workers in the state of New South Wales, yet the government felt free to fire over a thousand railway workers when it cut back on rail services. No one wept for the sacked railway workers, but crocodile tears were shed for loggers.

It is interesting that the only workers in the world who received good, sympathetic, almost laudatory coverage in the US press were those in Poland's Solidarity movement and in Russia. So communist workers are okay, but American workers are not. It all fits with the traditional hidden corporate agenda and the social engineering of America.

Interest rates, corporate managers, retail sales, trade deficits, and daily reports of the Dow Jones Industrial Average are of interest to only a very small minority of American people. (Although, at a time of severe financial turmoil that is affecting millions of people, there undoubtedly is more interest in the financial markets.) When I read the daily newspaper, it often seems so loaded with articles on finance, banks, profits, and corporate takeovers that I feel I'm reading a financial journal, when all I want is community, national, and international news. Perhaps it is all one and the same thing. After all, the corporations run Congress, the White House, the Pentagon, the media, the banks, and the Third World.

On that note, Rupert Murdoch is notorious for electing and dismissing national leaders. In 2003, he said to a congressional panel that the use of "political influence in our newspapers or television (is) nonsense." Before the Iraq war began in 2003, however, Murdoch openly stated that he fully supported and praised President Bush's decision to go to war in Iraq. Why? He wants cheaper oil. "The greatest thing to come out of this war for the economy . . . would be $20 a barrel for oil."[18] He owned five newspapers in Britain, and his media helped to elect Thatcher—they were her "cheer squad."[19] Murdoch also owns the *London Times*, the *New York Post*, and the *Wall Street Journal*. In Australia, where he owns more than 60 percent of the print media, he was a major supporter of former Prime Minister Bob Hawke. Murdoch makes a practice of interfering in the running of his newspapers, even after he has promised independence to his editors. When the story of torture at Abu Ghraib broke, the *New York Post* did not run it on the front page.[20]

The late Peter Jenkins, a respected journalist who once worked with Murdoch in Britain, wrote that promises of editorial freedom "are of very little weight against a proprietorial or managerial ethos which is unfriendly to honest, fair and decent professional journalism." He added, "I had no cause for personal complaint against Murdoch, but I saw how good newspapers, and once independent spirits, wilted in his presence—or at 3,000 miles removed."[21]

The standard of journalism these days is quite low, and I think that is because the basic ethic of the media currently is profit making. Excellent journalism has taken second place to infotainment and advertising space. *USA Today* epitomizes the "sound bite" mentality. Its articles are short, superficial, and lacking in substance; and they have lots of colored pictures, like comic books. Frequently, I am told by TV commentators that they do not have time to properly investigate a story because of timelines, but we depend on the media to educate and guide us so that we may save the Earth. Usually, I am given three minutes on the *Today Show* to tell the American people about the medical effects of nuclear war, while some film star is given eight minutes to discuss her latest face-lift. Important news is trivialized; and the rich, famous, or beautiful are worshiped. This sort of reporting insults the American public.

For a nation of its size, the United States has relatively few newspapers that offer sophisticated journalism. And even with these, one must learn to read between the lines and to think critically. I heard a former CIA agent openly admit in the International Court of Justice that the CIA places "disinformation" stories (overt lies about foreign affairs that then influence US political decisions) on the front page of the *New York Times*, the *Washington Post*, and other publications, as do the right-wing think tanks. Disinformation segments are also produced for TV.

In addition, something very serious has happened since the 9/11 tragedy. The US media as a whole were co-opted into sycophantic followers of everything that the White House promoted, including initiation of the war in Afghanistan (a war, incidentally, which will not and can never be "won"), and the ensuing war in Iraq. The rationale of the latter war had absolutely no associated relevance to the 9/11 attack, as events and history have quite plainly revealed. It seems, however, that the media as a whole was too terrified to contradict or to disagree with the president and all that he stood for at that time—patriotism, nationalism, and "might is right."

Reporters from newspapers and networks allowed themselves to be "embedded" with the troops and military in Iraq, thereby making impossible any sort of impartial reporting. Furthermore, because of the negative impact that honest reporting had created during the Vietnam War (when the US public was shown images of people being shot, napalmed, and dying), the US media passively complied with the Pentagon edict that no "nasty" images of people dying would be shown to the "sensitive" American public, for fear that then people would not automatically support this immoral and illegal war. So strict was this subservient, self-imposed censorship that the media even allowed themselves to be co-opted into the Bush embargo of photos of dead soldiers arriving back in the United States, an embargo recently reversed by President Obama.

Even some of the most distinguished newspapers actually sold their souls. The *New York Times* printed at least five front-page, influential articles by the journalist Judith Miller—who had received her information from such sources as Ahmed Chalabi (a discredited refugee from Iraq) and the now convicted former White House staffer Scooter Libby—supporting the invasion of Iraq. The *Times* subsequently issued a halfhearted apology, written by the ombudsman, that was deftly hidden near the back of the paper.

The Pentagon aptly recruited former military men, denoted as "military analysts"—most of whom were employed by large and highly influential weapons manufacturers—as spokespeople for Secretary of Defense Donald Rumsfeld's and Vice President Dick Cheney's Iraq war strategy. These people were flown at public expense for carefully orchestrated visits to Guantánamo Bay to verify that all was perfectly fine at that torture center, and they were then transported to Iraq to report on how the well the war was proceeding. At no stage did the broadcasting networks question the financial interests and corporate alliances of these former high-ranking military personnel.

Members of this group duly echoed the administration talking points, often having received detailed briefings about what they should say. Internal Pentagon documents called these so-called analysts "message force multipliers" who could be counted on to parrot the administration's "themes and messages" to millions of Americans and to obey the instruction not to quote their Pentagon sources. These so-called analysts, though treated with great respect during their frequent media appearances, were not actually journalists, but military men pushing the Bush agenda while simultaneously standing to make large profits for their respective military-industrial corporations.

At the same time, the administration also distributed hundreds of fake news segments to local TV stations, replete with fawning administration accomplishments, and they made covert payments to the Iraq media so that they would publish coalition propaganda.[22]

Finally, much was revealed in Scott McClellan's 2008 book *What Happened: Inside the Bush White House and Washington's Culture of Deception.* The former White House press secretary, who served in that post from July 2003 to April 2006, wrote that Bush had relied on propaganda to sell the war, that the press corps had been too easy on the administration during the run up to the war, and that some of McClellan's own assertions from the briefing room podium had been "badly misguided."[23] This book is extremely significant as Bush fades from the national scene because McClellan is one of the first White House insider acolytes to publicly tell the truth.

The US media now reach into almost every country on Earth, through satellite TV, videotapes, printed magazines, the Internet, and newspapers. But the culture of Hollywood is not applicable to the people of Fiji, New Guinea, or Africa. These populations see ads for Pepsi or Coke, with their messages of affluence, complete with cars and the "good life," and they want it. They listen to music and lyrics, and their culture is degraded by comparison with the glitz. The whole world is becoming "deculturalized" into a uniform "Coca-Cola society," wanting and needing an American way of life.[24] This is a terribly dangerous development because if 6.72 billion people lived the way the inhabitants of Hollywood do, the Earth would be destroyed within the next fifty to a hundred years. Remember that the typical US citizen pollutes twenty to a hundred times more than the average Third World person does. The Earth does not have the resources to sustain 6.72 billion people in affluence.

I wrote these words in 1991, and unfortunately this prophesy has come to pass. China with 1.3 billion people and India with 1.1 billion people are rapidly becoming affluent. Their populations now want millions of cars, televisions, refrigerators, big houses, and more and more electricity—emulating the rich Western world. And why shouldn't they have what we have?

The Australian Broadcasting Corporation (another ABC) is an autonomous body funded by the Australian government—answerable to no one, with no corporate sponsorship, similar in structure to the British Broadcasting Corporation (BBC). It is staffed by excellent investigative journalists, and one of the main reasons I like to come to home from the United States is to be able to tune in to the independent ABC.

Obviously, if the world is to survive, the press must not be used as a profit-making venture for a few people who are rich beyond compare and who now almost control the world. The place to start breaking down the corporate structure is the United States. If Americans all used their democracy appropriately, they could force members of Congress, who legally represent them, to legislate against private media ownership. Until they use the laws available at their disposal, they will be controlled by transnational corporations, which tell them only what they want them to know and which will amuse and numb them with trivia and superficial "entertainment"—violence, sex, and sport. I think all the major media—TV, newspapers, and radio—should be publicly owned but operated along the lines of the ABC of Australia and the BBC.

As Walter Cronkite put it in a spontaneous outpouring following the election of George H. W. Bush,

> We know that Thomas Jefferson was right when he said, "A democracy cannot be both ignorant and free!" We know that no one should tell a woman she has to bear an unwanted child. We know that religious beliefs cannot define patriotism.
>
> We know that it is not how one's lips are formed, but what they say. And we know there is freedom to disagree with all or part of what I've just said.
>
> But God Almighty, we've got to shout these truths in which we believe from the rooftops.[25]

11 Healing the Planet: Love, Learn, Live, and Legislate

The only cure is love. I just walked around my garden. It is a sunny, fall day and white fleecy clouds are scudding across a clear, blue sky. The air is fresh and clear with no taint of chemical smells, and the mountains in the distance are ringed by shining silver clouds. Earlier I picked a pan full of ripe cherry guavas to make jam, and the house is filling with the delicate aroma of simmering guavas. Figs are ripening on the trees and developing that gorgeous, deep red glow at the apex of the fruit. Huge, orange-colored lemons hang from the citrus trees, and lettuces, beetroots, and cabbages are growing in the vegetable garden. The fruits and vegetables are organically grown, and it feels wonderful to eat food that is free of human-made chemicals and poisons.

It is clear to me that unless we connect directly with the Earth, we will not have the faintest clue why we should save it. We need to have dirt under our fingernails and to experience that deep, aching sense of physical tiredness after a day's labor in the garden to really understand nature. To feel the pulse of life, we need to spend days hiking in forests surrounded by millions of invisible insects and thousands of birds and the wonder of evolution. Of course, I realize that I am very fortunate indeed to be able to experience the fullness of nature so directly—literally in my own backyard. For many people—especially those living in urban areas who are unable to travel out of them regularly—such an experience is difficult to come by. Still, I urge all to try in some way to make a direct connection with the natural world.

Only if we understand the beauty of nature will we love it, and only if we become alerted to learn about the planet's disease processes can we decide to live our lives with a proper sense of ecological responsibility. And finally, only if we love nature, learn about its ills, and live accordingly will we be inspired to participate in needed legislative activities to save the Earth. So my prescription for action to save the planet is this: Love, learn, live, and legislate.

We must, then, with dedication and commitment, study the harm we humans have imposed upon our beloved Earth. But this is not enough. The etiology of the disease processes that beset the Earth is a by-product of the collective human psyche and of the dynamics of society, communities, governments, and corporations that result from the human condition.

We have become addicted to our way of life and to our way of thinking. We must drive our cars, use our clothes dryers, smoke our cigarettes, drink our alcohol, earn a profit, look good, behave in a socially acceptable fashion, and never speak out of turn or speak the truth, for fear of rejection. The problem with addicted people, communities, corporations, or countries is that they tend to lie, cheat, or steal to get their "fix." Corporations are addicted to profit and governments to power, and as Henry Kissinger once said, "Power is the ultimate aphrodisiac."

The only way to break addictive behavior is to love and cherish something more than your addiction. When a mother and a father look into the eyes of their newborn baby, do they need a glass of beer or a cigarette to make them feel better? When you smell a rose or a gardenia, do you think of work or do you forget for a brief, blissful moment everything but the perfection of the flower? When you see the dogwood flowers hovering like butterflies among the fresh green leaves of spring, do you forget your worries?

Now, try to imagine your life without healthy babies, perfect roses, and dogwoods in spring. It would seem meaningless. We take the perfection of nature for granted, but if we woke up one morning and found all the trees dying, the grass brown, and the temperature 120°F, and if we couldn't venture outside because the sun would cause severe skin burns, we would recognize what we once had but didn't treasure enough to save. To use a medical analogy: we don't really treasure our good health until we lose it or experience a dreadful accident. When I am injured, I always try, immediately after the trauma, to psychologically recapture the moment before, when I was intact and healthy. But it is too late. It is not too late, though, for our planet. We have ten years of work to do, and we must start now. If we do not, it may be too late for the survival of most species, including, possibly, our own. "But what can I do?" I hear you ask. Let me tell you.

In the industrialized world, and indeed in most of the Third World now, governments are increasingly run and organized by a few corporations. But the corporatization of government is not conducive to global survival, as we have seen. A corporate mentality encourages greed, selfishness, and consumerism—not compassion for people or for nature. How, then, do we, the so-called little people, or the grassroots, destroy this corporate stronghold of our country's politicians and governments so that we can develop a kind, compassionate society?

I have some suggestions based on the Australian experience. The Australian system has many flaws that need correction, but in several important ways it provides a good model for the United States.

Compulsory Voting

In Australia, voting is compulsory—if people do not vote, they are fined $20 and have twenty-one days to pay it. I always get a feeling of pride as I stand at the polling booth watching my compatriots—old and young, rich and poor, fat and thin, black and white, healthy and infirm—step up to the booth to vote. They do it with a sense of responsibility and dignity that befits a healthy democracy. Mandatory voting invokes responsibility—knowing that they have to vote, people have thought carefully about what they are voting for.

In many US elections, fewer than 50 percent of eligible citizens vote. In the 2008 elections, however, an unusual percentage of the US population (aged 18 and older) voted. Until this presidential election, I believed that most Americans felt that they could have little, if any, influence on their rather corrupt politicians and system of government. They seemed to typically think, "What difference will my vote make?" But President Obama worked from the grassroots up and composed a campaign of millions of ordinary people, including millions of young people. This campaign was a wonderful boost to the American spirit, and for the first time in many decades people really felt that they could make a difference to the political direction that their country pursued.

In Australia, voter registration is compulsory, which enables compulsory voting. In the United States, registration has been made difficult for the average person. Rules and laws pertaining to voter registration differ from city to city and state to state. Poor people, especially, find it hard to travel to the place of registration in many locations because of inadequate public transportation, let alone the difficulty in deciphering the numerous forms and complying with the various regulations and conditions.

The historical reason for this grossly unfair system is clear: to limit the franchise. In order to eliminate the power and privileges of the wealthy and corporate elite, the United States needs compulsory registration and compulsory voting so that the political will and needs of all the people will be represented. There is a simple and cheap way to achieve uniform voter registration in the United States. Since the address of every citizen is known by the US Postal Service, registration could be effected by governmental decree.

Campaign Funding

I strongly suggest that political campaigns be completely funded by the federal government. When I ran for the Australian

Parliament, in March 1990, I received ninety-one cents from the government for every vote I got. I spent a total of $40,000, raised $13,000 from individual donations from my electorate, and contributed $13,000 from my own funds; and the government reimbursed me $13,000 after the election.

It is vital to pass legislation that prohibits funding of political campaigns by special-interest groups and corporations, because by the time most politicians are elected in the United States, they have prostituted themselves to the organizations that financed their extremely expensive campaigns. According to the Federal Election Commission (FEC), candidates who spent the most money in the 2006 elections won 92 percent of the time. That really says it all. Winners raised an average of $1.06 million; losers, an average of $304,000. And all this fund-raising seems crazy when you consider that each party in the 2006 congressional primary races featured only one candidate, so the voters had no real choice. The average senator running for election begins a campaign with about $1.43 million in the kitty, and raises on average an extra $6.2 million. Between 1999 and 2004, Nancy Pelosi, Speaker of the House of Representatives, raised $7.8 million for her candidacy.[1]

All this means that these people are forever running for office and perpetually raising funds, which takes them away from the legislative activities for which they were elected. This extraordinary process obviously excludes from Congress any person who is not a millionaire or who is not a gifted fund-raiser. In this situation, ordinary people can never participate in the governance of their country.

Old-Style Campaigning

Politics is not really politics anymore. It is run, for the most part, by Madison Avenue advertising firms, who sell politicians to the public the way they sell bars of soap or cans of beer. The campaigns are monitored by polls, and the whole event takes on a contrived quality. Aspiring politicians promote their images by appearing for thirty-second "sound bite" TV slots, surrounded by an attractive, socially acceptable family, smiling a big smile while mouthing inane phrases. Or the ads viciously attack the opponent, often by means of lies, slander, and other dirty tactics. The outrageous expense of these TV ads helps account for the high cost of political campaigns.

My political campaign lasted three weeks. It began when I had lunch with the editor of the local newspaper, who tried to persuade me to run by saying that Parliament needed more women. I was not at all responsive to his line of reasoning, until he began to discuss the Gorbachev peace initiatives. He asked me, "What is Australia doing to support Gorbachev?" Then the light turned on. I realized that the previous twenty years of my life had been devoted to educating people about the medical consequences of nuclear war. I saw that this path led logically and inexorably to running for Parliament in order to provide more effective support for the global peace movement.

I drove home from the luncheon with a sense of excitement and anticipation. I campaigned in my rural electoral district of seventy thousand by holding public meetings in all the towns and cities. People were very responsive, and they raised many intelligent and thoughtful questions. We discussed and debated various issues, and we all learned from the experience.

As it happened, I lost the election by 684 out of seventy thousand votes, but in the process, because of the policy of preferential voting, I defeated the incumbent, who was the leader of the National Party—the most conservative political party in Parliament—but my votes then were allocated to the next in line. The political pundits received a terrible shock, and Australia has not been the same since.

I urge a return to old-style campaigning in the United States—one in which politicians actually enter into a dialogue with their electorate, in which political issues are debated, and in which subjects affecting the fate of the Earth are seriously discussed. Eliminate TV and radio advertising, and stop politicians from condescending to and staying aloof from the public, so that ordinary people cease feeling ostracized and can have input into their government.

Never forget that your politicians are not your leaders—not even the president. You are their leaders, and they are your representatives.

Proportional Representation

The next step in the formation of a truly representative government is proportional representation. Women make up 50.7 percent of the US population, yet they have relatively little power or authority in the political system. The future of the world is in grave danger, and men continue to make the decisions, most of which have little relevance to saving the planet. I venture to suggest that if the composition of Congress were altered by a constitutional amendment mandating that half of the members of Congress be women, the world would be a very different place. Of course, the 50 percent would include women good and bad, stable and neurotic, smart and not so smart—as does the average mix of male politicians who now dominate Congress.

As we have seen, women also account for about half of the Earth's population, yet they do two-thirds of the work. This toil and responsibility has gained them virtually no political power. Those few, like Margaret Thatcher and Indira Gandhi, who do rise to the top have exercised the most powerful and ruthless behavior. Women who maintain their intrinsic feminine qualities are the women that the world desperately needs. Women must also give credit to their own intellectual abilities and realize that they are at least as intelligent as most men. We need to assume our place in the political sun and take over the job of helping to steer the planet toward a safe future for our children.

We must all develop the sense that we are an integral part of our governmental structure. There are many ways to achieve this sense. For instance, a young mother with three children and a part-time job will have little time to be deeply involved in political campaigns. But then again, she needs to guarantee her children's future. So she can join an organization that I founded in 1980, called Women's Action for Nuclear Disarmament (WAND), now called Women's Action for New Directions, which is one of the most effective lobbying bodies in Congress against nuclear weapons production. Our members are so well versed in the subject that even the most powerful hawks in Congress consult them before they draft legislation concerning nuclear weapons and delivery systems. Before WAND and the allied Professionals' Coalition for Nuclear Disarmament (composed of physicians, lawyers, educators, and scientists) and other NGOs, only the military-industrial organizations and the Pentagon helped set the legislative agenda. WAND has chapters in many towns and cities, so you can either join an established chapter or start one yourself. You can find WAND on the Internet at www.wand.org. WAND will help you find out who your federal and state representatives are, what committees they belong to, how they vote, and how you can influence their voting.

As you come to understand the political process better, and as your children grow up and you have more spare time, you may decide to run for local, state, or federal posts yourself. If you are a support person and not necessarily a leader, you will develop political campaigns for other women and give them the sort of strategies and nurturing they need. In 1991 I wrote, "Let's set a goal of 50 percent representation by the year 2000." Clearly this goal has failed to be realized. In 2007, in fact, female representation in both houses of Congress was a mere 16 percent; by comparison, in Rwanda it was 49 percent, and in Sweden, 47 percent.[2] Now we need to reset this goal to be achieved by 2015. Never has the country or the Earth needed our voice and our wisdom so desperately as it does now.

WAND accepts men too. If you are a man who cares deeply about the Earth, you are very welcome to join WAND. It needs sensitive, compassionate men who will transform political parties by their loving behavior and who are courageous enough to speak the truth with a passion rarely heard in the corridors of political power or, for that matter, in those of corporate power.

Politicians and Their Electorates

In 1989 I visited Iceland as a guest speaker at the fiftieth-anniversary commemoration of women's voting rights. Iceland is a fascinating, quite isolated country. Geologically, it is a volcanic island composed almost solely of black basalt rock, which supports very little life except moss, lichens, and tundra. Hot springs bubble from beneath the rocks, volcanoes hiss and spurt lava, and geysers soar majestically into the sky. Two massive tectonic plates meet in the middle of Iceland, and they are always moving, although ever so slowly. In such a place of planetary commotion, one gets the sense of a living, moving, evolving, magical Earth.

The politics of Iceland are less turbulent but just as fascinating. In the center of Reykjavík, the capital city, the prime minister and politicians congregate at lunchtime in the public saunas that are heated by underground hot springs. These eminent people sit in the baths alongside people from the general public, old and young, and conduct daily political conversations and dialogues. All issues are discussed, and the politicians return to work after lunch having been briefed by their public. Unfortunately the Icelandic economy has been devastated by the current global economic crisis.

Had I won office in March 1990 and entered the Australian Parliament, I would have taken my political directives from my electorate. I would have flown home every weekend from Canberra, the capital city, and conducted a public meeting in a different town or city on each occasion. I would have reported to my constituents about the preceding and following week's legislative activities, given them my opinion, and asked for their input. After suitable discussion and debate, I would have returned to Parliament armed with a consensus to guide and direct my activities and voting. I believe this is the only fair and legitimate way that democracy should be conducted.

I ran as an independent candidate; I received no money from special-interest groups and was therefore answerable only to my electorate. The two major political parties in Australia are basically corrupt because they receive large corporate campaign contributions and because the system of party loyalty allows no dissent or independence on the part of individual representatives. All members become a cog in the party machine, which is run by several tough fellows who metaphorically kick in heads if people do not conform.

In the United States, too, it is abundantly clear that most, though not all, politicians do not have the luxury or freedom to truly represent the wishes of their constituents, because they have already sold out almost before they are elected. Furthermore, both the Democrats and the Republicans are, for the most part, fundamentally conservative. As Gore Vidal once said, "America has one political party with two right wings."

It has by now become obvious that the politics of the past and present are no longer appropriate models for the politics of the future, if the Earth is to survive. I therefore propose that independent green candidates run in every congressional district and for every US Senate seat in future elections. Some will almost surely win, and even if just three or four independents enter the House of Representatives, the potential for change in the political system of the United States will be very real. The independents by their very presence will cause a sea change in the thinking and behavior of the Democratic and Republican parties. At the very least, the urgency of the task to rescue the planet will be a part of the official agenda. No longer will CFC legislation compromise with the fate of the Earth or with global warming; no longer will nuclear reactors be allowed to operate, producing hideous toxins to poison future generations; and no longer will corporate America be allowed to interfere with government agendas. Independent greens should also run for local and state government, a signal that the citizens of the United States have decided to become healers of the planet in their own right.

Elections will be fought and won in the public arena, organized with town hall meetings and political debates, door knocking, outdoor rallies, and energetic performances by the candidates. Huge political campaign funds will be rendered unnecessary once the candidates meet their constituents face-to-face instead of on television. The politics of the United States will then become truly democratic again, in the style of Lincoln and Jefferson.

I hesitate at this stage to suggest the formation of a national green party, because organizations of any sort tend to begin with revolutionary, fresh thinking but, with growing success, tend to become mired in bureaucracy and power struggles. This is old-style politics. Rather, I see the candidate and the electorate as forming a separate, autonomous unit that takes its seat in the House or the Senate and that votes as its own bloc, free of external influences.

With such an arrangement, we will really have representational democracy.

This is not a pipe dream. The vote for independent green candidates is on the increase in Australia and indeed worldwide, and the United States needs to catch up with this new ecological and political trend. There is great hope and excitement in this development, and I encourage you to become the candidate of the future in your congressional district. It is time for the second American Revolution. If Soviet citizens can rebel, so can Americans.

Finally, I am pleased to announce a specific and exciting "roadmap" that will provide the United States of America with an energy future that produces no carbon dioxide and does not utilize nuclear electricity, by the year 2050. This roadmap lays out directives for the US public, corporations, and politicians to follow in order to avert the full and catastrophic impacts of global warming. Titled *Carbon-Free and Nuclear-Free: A Roadmap for U.S. Energy Policy*, and compiled by a brilliant scientist, Arjun Makhijani, it outlines the pragmatic economic steps necessary to achieve these goals, which in the long run will change little in the standard of living in the US.[3]

Briefly, a rapid introduction of renewable energies that are technologically and currently available includes the following:

1. *Wind*. The potential for wind energy in the United States is vast; wind energy can provide over two and a half times the total electricity generation that was required in the United States in 2005. In fact, the wind energy potential in each of the top six states—North Dakota, Texas, Kansas, South Dakota, Montana, and Nebraska—is greater than the total nuclear electricity production from the 103 operating US nuclear reactors. Unfortunately, because the wind supply is located in states that tend to be far from the major grids, there is an urgent need to upgrade the national grid. Farmers are making more money generating electricity than growing food, but they can do both on the same land. Denmark, Spain, and Germany now generate up to 20 percent of their electricity from wind.[4]

2. *Solar.* Solar power, including power generated by solar photovoltaics and solar thermal power plants, has enormous potential. At 20 percent efficiency and only 1 percent of the US land area, the potential for solar electricity generated by photovoltaic cells is about eight times the total US electricity production. Until now, solar power has been less economical than wind, but with new developments that situation is rapidly changing.[5] For example, plans for two massive solar plants are under way for California to produce a total of 800 megawatts, equivalent to the electricity generated from a large coal or a small nuclear power plant. Solar energy in the form of biomass, including biofuels derived from solar energy, was discussed in Chapter 3. Direct hydrogen production can be developed from solar energy.[6]
3. *Geothermal.* Geothermal, or hot rock, energy uses the natural heat deep underground to generate electricity. The advantage there is that the supply is not intermittent, as with wind or solar, so geothermal sources can produce baseload power.[7]
4. *Wave energy.* Wave energy is steady and reliable—less intermittent than wind. The Electric Power Research Institute estimated that Hawaii, Oregon, northern California, and Massachusetts could achieve wave power on an economic par with wind energy.[8] Tidal power utilizes the twice-daily changes in sea level to generate electricity. It is suitable only for certain coastlines, but it certainly offers great possibilities in places where the tides vary 20–100 feet (6–30 meters) each day. Wave power is another dynamic area awaiting development.
5. *Cogeneration.* Cogeneration is a wonderful method for harnessing heat, usually wasted in factories. One technique, used extensively in Russia and in Scandinavia, is to heat water and pump it to warm whole towns and cities. Another is to use waste steam to drive electricity-generating turbines, to run refrigerators, and to power industrial machinery. An ordinary power plant is 32 percent efficient, but a cogenerator consuming the same amount of fuel is 80 percent efficient.[9]

6. *Conservation*. Conservation can save large quantities of energy. Society must invest in highly efficient lightbulbs, refrigerators, stoves, cars, and street lighting. Energy-efficient equipment uses one-third to one-half less energy than does conventional technology. Much of it has already been invented, but monopolistic corporations tend to encourage distribution of inefficient equipment, thus leading to increased electricity consumption. For example, General Electric manufactures not only nuclear reactors but also hair dryers, toasters, stoves, clothes dryers, and refrigerators. Is it not therefore in GE's best interest to encourage people to use more electricity with less efficient appliances and to use electric brooms, electric hedge clippers, and electric lawn mowers instead of ones operated by healthy muscle power?

The truth is, energy-efficient investments are much cheaper financially and ecologically than the building and operating of coal or nuclear plants. Patents for wonderful energy-saving inventions abound, but most inventors lack the money to develop their products. And corporations seem uninterested in pursuing or financing such inventions.

Not least, utilities hide enormous government subsidies that they receive for fossil fuels and nuclear power. This deception makes renewable energy appear more expensive. Because utilities enjoy an almost total monopoly in energy advertising and technologies of energy production, it is very hard to understand and dissect their propaganda. Solar heating systems and photovoltaic cells endow people with energy self-reliance, but clearly such self-sufficiency is not and will not be seen to be in the best interest of the utilities.

According to Makhijani, the first four in his list—wind, solar, geothermal, and wave power—could theoretically supply the entire 2050 US energy requirement. Here is the prescription for our survival!

We all have powers and talents that are unique and that the Earth needs. We no longer have the right to hide our light under a bushel; the world needs our God-given talents—whatever they may be—and if we all decide to become autonomous and powerful in our own right, we will save the Earth.

Remember, though, that we must start now—this instant—for there is no time to waste. We have less than the next ten years, and each moment is precious. Always stay in the light, always be hopeful, and if obstacles arise, step over them, through them, or under them. If you have a sense of destiny and such knowledge that good will prevail over evil, we will save the Earth.

Hope for the Earth lies not with leaders, but in your own heart and soul. If you decide to save the Earth, it will be saved. Each person can be as powerful as the most powerful person who ever lived—and that is you, if you love this planet.

Appendix: Resources for More Information

What follows is a listing of some organizations and media that focus on the issues presented in this book. In many cases, the organization's own words are used in the descriptions.

Helen Caldicott, MD

DR. CALDICOTT'S PERSONAL WEBSITE

www.helencaldicott.com

IF YOU LOVE THIS PLANET

Dr. Helen Caldicott's weekly, one-hour radio program about environmental, nuclear, and other public health issues. Website has archive of all shows, news articles, and studies.

www.ifyoulovethisplanet.org

BEYOND NUCLEAR

Washington DC–based group dedicated to educating the public about the connection between nuclear power and nuclear weapons, and the need to abandon both to safeguard the future. Teaches that nuclear energy is not a solution to global warming, and illuminates the dangers of nuclear waste.

www.beyondnuclear.org

INSTITUTE FOR ENERGY AND ENVIRONMENTAL RESEARCH (IEER)

Dedicated to increasing public involvement in and control over environmental problems through the democratization of science. Arjun Makhijani, PhD, president, wrote the groundbreaking book *Carbon-Free and Nuclear-Free: A Roadmap for U. S. Energy Policy,* a blueprint for achieving a pollution-free future by 2050, which can be downloaded on the IEER website. Fiscal sponsor of Dr. Caldicott's radio show, *If You Love This Planet* (www.ifyoulovethisplanet.org).

wwwieer.org

PEOPLE FOR A NUCLEAR-FREE AUSTRALIA (ALSO DOCTORS FOR A NUCLEAR-FREE AUSTRALIA)

Australian organization that opposes uranium mining, the shipping of uranium to other countries for use in overseas reactors (which would increase radioactive waste), and the building of any nuclear reactors in Australia.

www.pnfa.org.au

State of the Planet

EARTH POLICY INSTITUTE (EPI)

Think tank dedicated to building a sustainable future, an "economy for the Earth," as well as providing a plan of how to get from here to there. Founder Lester Brown established the Worldwatch Institute. Principal research areas include food, population, water, climate change, and renewable energy. Offers a free newsletter: *Earth Policy News*.

www.earth-policy.org

FORUM FOR THE FUTURE

Organization that is committed to sustainable development and developing an enriching and enjoyable way of life that does not threaten the global environment or rob future generations of resources. Projects include creating a low-carbon economy, tracking sustainable cities, and a global competition paying $75,000 for innovative approaches, products, or services that address climate change.

www.forumforthefuture.org

MILLENNIUM ECOSYSTEM ASSESSMENT

UN reports issued after 2001 that are intended to assess the consequences of ecosystem change for human well-being and the scientific basis for

action needed to enhance the conservation and sustainable use of those systems. Issues covered include health, biodiversity, business challenges and opportunities, desertification, and wetlands.
www.millenniumassessment.org

UNITED NATIONS ENVIRONMENT PROGRAMME (UNEP)

Arm of the United Nations that provides leadership and encourages partnership in caring for the environment by inspiring, informing, and enabling nations and peoples to improve their quality of life without compromising that of future generations. Issues include energy, biodiversity, regional seas, ozone, conflict and disaster, species, chemicals, indigenous peoples, and sustainable consumption.
www.unep.org

WORLD RESOURCES INSTITUTE (WRI)

Environmental think tank seeking practical ways to protect the Earth and improve people's lives, with four topic areas: (1) climate, energy, and transport; (2) governance and access; (3) markets and enterprise; (4) people and ecosystems.
www.wri.org

WORLDWATCH INSTITUTE

Independent research organization that analyzes critical global issues and intends to empower decision makers to build an ecologically sustainable society that meets human needs. Publishes bimonthly print magazine *World Watch*, as well as annual reports such as *State of the World* and *Vital Signs*. Two recent *State of the World* reports are subtitled *Our Urban Future* and *China and India*.
www.worldwatch.org

General Environmental Organizations

BIONEERS

Nonprofit organization that promotes practical environmental solutions and innovative social strategies for restoring Earth's imperiled ecosystems and healing our human communities. Sponsors annual conference in the San Francisco area, open to the public.
www.bioneers.org

CONSERVATION INTERNATIONAL

International conservation group that believes that the Earth's living heritage—its global biodiversity—must be maintained if future generations are to thrive spiritually, culturally, and economically, and to live harmoniously with nature. Issues include saving species, conserving landscapes and seascapes, and empowering local communities to ensure responsible use of natural resources.
www.conservation.org

EARTHJUSTICE

Nonprofit public-interest law firm that works to protect the magnificent places, natural resources, and wildlife of this Earth, and to defend the right of all people to a healthy environment—"because the Earth needs a good lawyer." Brings about far-reaching change by enforcing and strengthening environmental laws on behalf of hundreds of organizations, coalitions, and communities. Major issue areas are global warming, air, forests, health and communities, oceans, public lands, water, and wildlife. Releases an annual issue paper on the status of environmental rights around the world.
www.earthjustice.org

ENVIRONMENT AMERICA

Federation of state-based, citizen-funded environmental advocacy organizations. Combines independent research, practical ideas, and tough-minded advocacy to overcome the opposition of powerful special-interest groups and win real results for the environment. Issues include promoting solar and wind power, cutting emissions that cause global warming and toxic pollution, and fighting for healthy oceans, parks, wild spaces, rivers, lakes, and streams.
www.environmentamerica.org

ENVIRONMENTAL DEFENSE FUND (EDF)

Environmental advocacy group that, since 1967, has linked science, economics, law, and innovative private-sector partnerships to create breakthrough solutions to the most serious environmental problems.
www.edf.org

FRIENDS OF THE EARTH (FOE)

Environmental organization that protects the rights of all people to live in a safe and healthy environment, with campaigns demonstrating the

belief that the fight for justice and the movement to protect the health of the planet are part of the same struggle.
www.foe.org

GLOBAL GREEN

American arm of Green Cross International, which was created by President Mikhail Gorbachev to foster a global value shift toward a sustainable and secure future by reconnecting humanity with the environment. In the United States, focused on stemming global climate change by creating green buildings and cities. Internationally, works to eliminate weapons of mass destruction that threaten lives and the environment, and to provide clean, safe drinking water for the 2.4 billion people who lack access to clean water.
www.globalgreen.org

GREEN FESTIVAL

The largest sustainability event in the world, held annually in several US cities. Includes speakers and exhibitors, and is sponsored by Global Exchange and Green America (formerly Co-op America).
www.greenfestivals.org

GREENPEACE

International organization focused on environmental and nuclear issues, including global warming, destruction of ancient forests, deterioration of our oceans, and the threat of a nuclear disaster. Has offices in several major cities around the world. Website has a great deal of useful data about all major environmental issues, and regularly updated news and information on campaigns.
www.greenpeace.org

HARVARD MEDICAL SCHOOL CENTER FOR HEALTH AND THE GLOBAL ENVIRONMENT

Organization that focuses on expanding environmental education at medical schools and investigating and promoting awareness of the human health consequences of global environmental change. The center, the only one of its kind at a US medical school, focuses on making the strongest possible case that human beings are an intimate part of the environment and that we cannot damage it without damaging ourselves. Co-produced the 2005 report Climate Change Futures: Health, Eco-

logical and Economic Dimensions (www.climatechangefutures.org/pdf/CCF_Report_Final_10.27.pdf).
http://chge.med.harvard.edu

LEAGUE OF CONSERVATION VOTERS (LCV)
Effective environmental advocacy group that works to secure the environmental future of our planet by advocating for sound environmental policies and for electing pro-environment candidates who will adopt and implement such policies. Through its "National Environmental Scorecard" and "Presidential Report Card," informs the public about the most important environmental legislation of the past congressional session and shows them how their own and others' representatives voted.
www.lcv.org

NATURAL RESOURCES DEFENSE COUNCIL (NRDC)
Environmental action group that uses law, science, and member/activist support to protect the planet's wildlife and wild places and to ensure a safe and healthy environment for all living things. Issues include curbing global warming, reducing toxic chemicals, moving beyond oil, reviving our oceans, and helping China go green.
www.nrdc.org

NATURE CONSERVANCY
The leading conservation organization working around the world to protect ecologically important lands and waters for nature and people. Issues include climate change (including rising sea levels), forests, grasslands, and deserts. Has launched the Campaign for a Sustainable Planet—what it calls the largest conservation campaign in history—"to ensure the survival of the natural world that sustains us all."
www.nature.org

PHYSICIANS FOR SOCIAL RESPONSIBILITY (PSR)
Medical and public health organization advocating policies to stop nuclear war and proliferation and to slow, stop, and reverse global warming and toxic degradation of the environment. Addresses environmental topics such as coal, toxic chemicals, climate change and human health, and the hazards of nuclear power. Has thirty-one chapters in the United States.
www.psr.org

PUBLIC EMPLOYEES FOR ENVIRONMENTAL RESPONSIBILITY (PEER)

Service organization assisting federal and state public employees, enabling public servants to work as "anonymous activists" so that agencies must confront the message, rather than the messenger.

www.peer.org

SIERRA CLUB

"America's oldest, largest and most influential grassroots environmental organization." Inspired by nature, its members work together to protect their communities and the planet. Campaigns focus on urban sprawl, wildlands, forest protection and restoration, clean water, global warming, human rights and the environment, global population, clean air, species and habitat, nuclear waste, sustainable consumption, toxics, and wetlands.

www.sierraclub.org

TRANSITION TO GREEN

Environmental transition recommendations for the Obama administration. In December 2008, twenty-nine of the nation's major green groups, representing a spectrum of opinion, published a 391-page document with a recommended environmental action plan for then President-elect Barack Obama. *Transition to Green*, subtitled *Leading the Way to a Healthy Environment, a Green Economy and a Sustainable Future*, calls for Obama administration policies that combine environmental, energy, and economic needs. Urges Congress to pass legislation in 2009 to cut emissions 35 percent below 2008 levels by 2020, and 80 percent below 1990 levels by mid-century. Also wants to see "energy efficiency, modernizing the grid, and greatly expanding power generation from renewable energy resources" to make real headway in achieving 100 percent clean electricity.

http://gristmill.grist.org/images/user/6337/transition_to_green_full_report.pdf

THE WILDERNESS SOCIETY

A leader in conservation and wilderness protection since 1935. Wrote and passed the landmark Wilderness Act, which won lasting protection for 107 million acres (about 43.3 million hectares) of wilderness, including 56 million acres (about 22.7 million hectares) of spectacular lands in Alaska, 8 million acres (about 3.2 million hectares) of fragile desert lands in California, and millions more throughout the nation. Calls for protecting America's wilderness, not as a relic of the nation's past but as a thriving ecological community that is central to life itself.

www.wilderness.org

WOMEN'S ACTION FOR NEW DIRECTIONS (WAND)

Progressive organization that was created to empower women to act politically to reduce violence and militarism, and redirect excessive military resources toward unmet human and environmental needs. Current priorities include reducing excessive military spending in order to achieve more balanced spending priorities, blocking new US nuclear weapons and reducing funding for missile defense, preventing reprocessing of nuclear waste, and supporting efforts to store nuclear waste in the least dangerous manner. Founded by Dr. Helen Caldicott in 1982.
www.wand.org

Ozone Depletion

NASA OZONE HOLE WATCH

Website that allows visitors to check on the latest status of the ozone layer over the South Pole. Satellite instruments monitor the ozone layer, and the data are used to create images that depict the amount of ozone.
http://ozonewatch.gsfc.nasa.gov

NOAA/NATIONAL WEATHER SERVICE CLIMATE PREDICTION CENTER: OZONE LAYER

Web page showing recent measurements of the ozone layer.
www.cpc.ncep.noaa.gov/products/stratosphere/sbuv2to/sbuv2to_latest.shtml

UNITED NATIONS ENVIRONMENT PROGRAMME, OZONACTION BRANCH

Important UN program that assists developing countries and countries with economies in transition (CEITs) to enable them to achieve and sustain compliance with the Montreal Protocol. Ozone news, reports, studies, and conferences.
www.uneptie.org/ozonAction

Global Warming

THE BBC WEATHER CENTRE: CLIMATE CHANGE

Page of the BBC's weather-forecasting service that provides basic information about the effects of global warming.
www.bbc.co.uk/climate

CLIMATE CHANGE DENIAL

Blog that explores the psychology of climate-change denial, with observations and anecdotes about the sometimes disturbed response to the problem. Ponders why—when the evidence is so strong, and so many agree that global warming is a major problem—we are doing so little about climate change.

www.climatedenial.org

THE CLIMATE PROJECT

Fan education program that focuses on increasing public awareness of the climate crisis at a grassroots level in the United States and abroad. One thousand volunteers trained by Al Gore have done nearly twenty thousand presentations based on Gore's film *An Inconvenient Truth* to teach people about global warming.

www.theclimateproject.org

11TH HOUR ACTION NETWORK

Action groups that use Leonardo DiCaprio's film *The 11th Hour* as a teaching tool to raise public awareness of global warming.

www.11thhouraction.com

AL GORE

Information about global warming educator Al Gore's projects and organizations. Website has educational materials.

www.algore.com

GREENPEACE: CLIMATE CHANGE CAMPAIGN

Greenpeace campaign that focuses on climate change. Website has news, action items, and reports to empower visitors to fight climate change, including stopping coal emissions.

www.greenpeace.org/international/campaigns/climate-change

JAMES E. HANSEN

Website of Professor James Hansen, for over twenty years a leading climate researcher, that publicizes the urgency of global warming. Contains Hansen's reports and statements.

www.columbia.edu/~jeh1

HEAT IS ONLINE

Education website focused on all aspects of global warming. Created by journalist/author Ross Gelbspan.

www.heatisonline.org

AN INCONVENIENT TRUTH

Official website of Al Gore's film *An Inconvenient Truth*, in which Gore presents an illustrated lecture about the effects of global warming.

www.climatecrisis.net

INTERGOVERNMENTAL PANEL ON CLIMATE CHANGE (IPCC)

Important international institution established to provide decision makers and others interested in climate change with an objective source of information. Published reports in 2007 and 2008 about the potential effects of global warming. Does not conduct research or monitor climate-related data or parameters.

www.ipcc.ch

LIVESTOCK'S LONG SHADOW: ENVIRONMENTAL ISSUES AND OPTIONS

2006 report by the Food and Agriculture Organization of the United Nations (FAO) that documents how the livestock sector generates more greenhouse gas emissions as measured in CO_2 equivalent—18 percent—than does transport, and is also a major source of land and water degradation.

ftp://ftp.fao.org/docrep/fao/010/a0701e/a0701e.pdf

MET OFFICE: CLIMATE CHANGE

Climate change page of Britain's national meteorological service—"one of the world's leading providers of environmental and weather-related services." Includes research into what could happen under climate change, including the impacts on current and future generations.

www.metoffice.gov.uk/climatechange

PEW CENTER ON GLOBAL CLIMATE CHANGE

US-based organization that brings together business leaders, policy makers, scientists, and other experts to construct a new approach to a complex and often controversial issue. Believes we can work together to protect the climate while sustaining economic growth.

www.pewclimate.org

REALCLIMATE.ORG

Commentary on climate science by working climate scientists for the interested public and journalists. Aims to provide a quick respose to developing stories and provide the context sometimes missing in mainstream commentary.

www.realclimate.org

ROYAL SOCIETY: CLIMATE CHANGE

Global-warming arm of the national academy of science of the United Kingdom and the Commonwealth, at the cutting edge of scientific progress.

www.royalsoc.ac.uk/climatechange

SCIENCE AND DEVELOPMENT NETWORK: CLIMATE CHANGE & ENERGY

Climate change and energy page of the Science and Development Network, focused on providing reliable and authoritative information about how science and technology can reduce poverty, improve health, and raise living standards in the developing world.

www.scidev.net/climate

STERN REVIEW: CONCLUSIONS

Conclusions of the 2006 report *Stern Review: The Economics of Climate Change*, written by economist Nicholas Stern for the British government, covering the effects of global warming on the world economy.

www.thefirstpost.co.uk/1910,features,-time-to-get-stern-on-climate-change,3

DAVID SUZUKI FOUNDATION

Canadian organization committed to balancing human needs with the Earth's ability to sustain all life. Strives to find and communicate practical ways to achieve that balance.

www.davidsuzuki.org

UNION OF CONCERNED SCIENTISTS: GLOBAL WARMING

Global-warming page of the Union of Concerned Scientists, a group of citizens and scientists working for environmental solutions. Various issues include global warming, energy, nuclear weapons, nuclear power, and species extinction.

www.ucsusa.org/global_warming

WE CAN SOLVE THE CLIMATE CRISIS

A project of the Alliance for Climate Protection, an effort founded by Al Gore. The alliance's goal is to create the political will to solve the climate crisis—in part through repowering the United States with 100 percent of its electricity from clean energy sources within ten years.
www.wecansolveit.org

Peak Oil

POST CARBON INSTITUTE

Timely organization which conducts research, educates the public, and organizes leaders to help communities around the world understand and respond to the challenges of fossil fuel depletion and climate change, primarily by meeting more of a community's needs from local sources, and less from faraway sources.
www.postcarbon.org

Oceans

GREENPEACE: DEFENDING OUR OCEANS CAMPAIGN

Active campaign that sets out to protect our global oceans now and for the future from exploitation and controllable human pressure, to allow them to recover. Educates the public that without the global ocean there would be no life on Earth, and that the oceans need all the resilience they can muster in the face of climate change.
www.greenpeace.org/international/campaigns/oceans

JOINT OCEAN COMMISSION INITIATIVE

Initiative seeking to expand our collective understanding of the threats facing our oceans and to enable actions that address those threats so that our oceans remain vibrant and healthy for current and future generations. Builds support for ocean policy reform at all levels of decision making.
www.jointoceancommission.org

NATIONAL CENTER FOR ECOLOGICAL ANALYSIS AND SYNTHESIS (NCEAS): MARINE ECOLOGY AND CONSERVATION

Scientific organization that addresses fundamental questions about ecological and evolutionary processes, and provides information to resource manage-

ment and conservation professionals. Hundreds of publications and presentations have been produced from NCEAS's work on marine systems.
www.nceas.ucsb.edu/ecology/marine

OCEAN ALLIANCE
Alliance dedicated to the conservation of whales and their ocean environment through research and education. Collaborates with some of the world's leading scientists, and has produced over thirty-four documentaries and won numerous awards and honors.
www.oceanalliance.org

OCEAN CONSERVANCY
Conservation group that promotes healthy and diverse ocean ecosystems and opposes practices that threaten ocean life and human life. Through research, education, and science-based advocacy, informs, inspires, and empowers people to speak and act on behalf of the oceans. Issues include climate change and the ocean (the ocean will be the first ecosystem to suffer widespread effects of global warming), entangled animals, coral reefs, marine debris, offshore drilling, overfishing, sustainable seafood, and whaling.
www.oceanconservancy.org

OCEANA
Large, international ocean protection and restoration environmental advocacy group dedicated to protecting and restoring the world's oceans. Concerns include pollution, and preventing the collapse of fish populations, marine mammals, and other sea life. Teaches that oceans cover 71 percent of the globe, and that they not only control our climate, but are also the primary source of protein for one billion people. Brought to light the fact that shipping now accounts for 3 percent of greenhouse gases (double the emissions of aviation), and that shipping's emissions might triple by 2030. The related article "Shipping's Impact on the Air" can be found at http://edition.cnn.com/2008/WORLD/asiapcf/01/20/eco.about.ships.
www.oceana.org

SCRIPPS INSTITUTION OF OCEANOGRAPHY
A leading global oceanographic research foundation led by ocean science researchers and students. Conducts extensive climate and atmospheric research, and studies fragile ecosystems. Found that sea level rise will

have different consequences in different places but that the rise will be profound on virtually all coastlines. Scripps scientists also discovered that nitrogen trifluoride is a major greenhouse gas. Relevant reports include "Potent Greenhouse Gas More Prevalent in Atmosphere Than Previously Assumed" (http://scrippsnews.ucsd.edu/Releases/?releaseID=933) and "Nitrogen Trifluoride in the Global Atmosphere" (www.agu.org/pubs/crossref/2008/2008GL035913.shtml).
www.scripps.ucsd.edu

SEAFLOW
Educational organization building an international movement dedicated to protecting whales, dolphins, and all other marine life from active naval sonars and other lethal ocean noise pollution. Draws on science, creative action, the arts, and community for inspired participation to safeguard the web of life.
www.seaflow.org

Water

BLUE PLANET PROJECT
Encompassing organization that intends to protect the world's freshwater from the growing threats of trade and privatization. Works with organizations and activists in both the south and the north using a human rights framework to maintain access to water for people and nature for generations to come. Building a movement to secure an international treaty on the right to water.
www.blueplanetproject.net

FOOD & WATER WATCH
Organization that works against the privatization of water—which is often cited as the "oil of the twenty-first century"—and for the removal of toxic chemicals and bacteria from the water supply.
www.foodandwaterwatch.org

Green Energy / Nuclear Energy Hazards

AMERICAN WIND ENERGY ASSOCIATION
The national trade association for the fast-growing wind-energy industry. Website has fact sheets, updated news, newsletter, and reports.
www.awea.org

BEYOND NUCLEAR

Washington DC–based group dedicated to educating the public about the connection between nuclear power and nuclear weapons, and the need to abandon both to safeguard the future. Teaches that nuclear energy is not a solution to global warming, and illuminates the dangers of nuclear waste.

www.beyondnuclear.org

BULLETIN OF THE ATOMIC SCIENTISTS

News and commentary about climate change, nuclear weapons, nuclear power, and biosecurity.

www.thebulletin.org

CARBON-FREE AND NUCLEAR-FREE: A ROADMAP FOR U.S.ENERGY POLICY

Groundbreaking report by engineer Arjun Makhijani, PhD, president fo the Institute for Energy and Environmental Research, which demonstrates how the United States can supply all of its energy needs by 2050 without carbon emissions or nuclear power. Downloaded over one million times since it was published in August 2007.

www.ieer.org/carbonfree

EV WORLD

News articles and resources about electric cars, present and future. Buyer's guides, audio and video clips, newsletter, chat group.

www.evworld.com

GREASECAR

Information and conversion kits for converting diesel automobiles and trucks to run on the Greasecar system. Greasecars use only straight vegetable oil, which does not have the toxic chemicals found in regular biodiesel and greatly reduces greenhouse emissions.

www.greasecar.com

HYDROGEN CAR

Informational website about all aspects of hydrogen cars.

www.hydrogencarsnow.com

INSTITUTE FOR ENERGY AND ENVIRONMENTAL RESEARCH (IEER)
Scientific organization dedicated to increasing public involvement in and control over environmental problems through the democratization of science. Arjun Makhijani, PhD, president, wrote the groundbreaking book *Carbon-Free and Nuclear-Free: A Roadmap for U.S. Energy Policy*—a blueprint for achieving a pollution-free future by 2050, which can be downloaded on the IEER website. Sponsors Dr. Caldicott's radio show *If You Love This Planet* (www.ifyoulovethisplanet.org).
www.ieer.org

NUCLEAR INFORMATION AND RESOURCE SERVICE (NIRS)
National information and networking center for citizens and environmental activists concerned about nuclear power, radioactive waste, radiation, and sustainable energy. Initiates large-scale organizing and public education campaigns on specific issues, such as preventing construction of new reactors, transportation of radioactive waste, deregulation of radioactive materials, and more.
www.nirs.org

NUKEFREE.ORG
Committed to preventing the construction of new nuclear reactors and to helping pave the way for an energy economy based on renewables, efficiency, and conservation. Website provides news on the most important nuclear power industry battles taking place across the country. Works closely with other groups monitoring energy issues. See "Debunking the Myths about Nuclear Power" (www.nukefree.org/sites/nukefree.org/files/debunk-nuke-myth.pdf).
www.nukefree.org

PROJECT LAUNDRY LIST
Nonprofit organization dedicated to making air-drying laundry acceptable and desirable as a simple and effective way to save energy. Dr. Caldicott is on the board of advisors.
http://laundrylist.org/

PUBLIC CITIZEN: ENERGY PROGRAM
Green-energy arm of Ralph Nader's public-interest, consumer advocacy organization Public Citizen.
www.citizen.org/cmep

REAL GOODS SOLAR, INC.

The foremost source of information and products in renewable energy, sustainable living, alternative transportation, and relocalization ("the ultimate anecdote [*sic*] for global warming and peak oil"). The oldest and largest catalog firm devoted to the sale and service of renewable energy products in the world.

www.realgoods.com

RENEWABLES 2007 GLOBAL STATUS REPORT

Worldwatch Institute report on the global renewable energy situation, and the growth of electricity, heat, and fuel production capacities from renewable energy sources, including solar photovoltaics, wind power, solar hot water/heating, biofuels, hydropower, and geothermal.

www.worldwatch.org/node/5630

ROCKY MOUNTAIN INSTITUTE

Institution founded by resource analysts L. Hunter Lovins and Amory B. Lovins, and focused on energy policy. Core principles include natural capitalism, positive solutions, and biological insight. See their page debunking negative myths about hydrogen cars at www.rmi.org/images/other/Energy/E03-05_20HydrogenMyths.pdf.

www.rmi.org

SOLAR LIVING INSTITUTE

Nonprofit educational organization whose mission is to promote sustainable living through inspirational environmental education, including hands-on workshops on renewable energy, green building, sustainable living, permaculture, organic gardening, and alternative (environmental) construction methods. Headquartered at the Solar Living Center in Hopland, California. Sponsors SolFest exposition each August.

www.solarliving.org

2020 VISION

Group dedicated to solving global challenges where international security, energy, and the environment come together. Contends that today's great global challenges are linked, and strives to find overarching solutions. To achieve its goals, works with Congress, with the media, on college campuses, and in communities across the country. Issues include oil addiction, global security, energy, and the environment.

www.2020vision.org

UNION OF CONCERNED SCIENTISTS

Group of citizens and scientists working for a healthy environment and a safer world. Combines independent scientific research and citizen action to develop innovative, practical solutions and to secure responsible changes in government policy, corporate practices, and consumer choices. Issue areas include global warming, clean energy, clean vehicles, safety concerns about nuclear power and nuclear weapons, food/agriculture, and invasive species.
www.ucsusa.org

WHO KILLED THE ELECTRIC CAR

Website that augments the documentary film about how General Motors' popular prototype electric cars were suddenly destroyed and never made available to consumers. Provides information about current and future green cars—hybrid, electric, hydrogen, etc.
www.sonyclassics.com/whokilledtheelectriccar

Green Lighting

ION & LIGHT COMPANY

Lighting company that distributes the entire line of fixtures from Chromalux, the original maker of full-spectrum lightbulbs. For more information about Chromalux bulbs, go to www.lumiram.com. Two relevant articles are "A Closer Look at Compact Fluorescents—The Environmental Benefit versus Health Impact" (http://vitalitymagazine.com/earthwatch_5) and "Fluorescent Light Ain't Right" (http://nyc.indymedia.org/en/2007/01/81805.html). To read several medical studies on the harmful health effects of fluorescent light, go to the "Reports, Studies and Conferences" page of www.ifyoulovethisplanet.org and look under the "Toxic Pollution" catagory.
www.ionlight.com

Trees / Rain Forests / Paper

AMAZON WATCH

Works with indigenous and environmental organizations in the Amazon Basin to defend the environment and advance indigenous peoples' rights in the face of large-scale industrial development—oil and gas pipelines, power lines, roads, and other megaprojects. Sponsors ChevronToxico

.com, the international campaign to hold Chevron Texaco accountable for its toxic contamination of the Ecuadorian Amazon.
www.amazonwatch.org

GLOBAL FOREST COALITION
Supports and coordinates joint NGO/IPO campaigns that focus on raising awareness of the need for socially just and effective forest policy, challenging the underlying causes of forest loss, and upholding the rights of indigenous and other forest peoples. Raises awareness of the social impacts of market-based conservation mechanisms, including carbon trading, gene trading, and the sale of environmental services in general. Fights the expansion of the agro-fuel business and opposes genetically modified trees.
wwwglobalforestcoalition.org

ILOVEMOUNTAINS.ORG
Action and resource center to end mountaintop removal coal-mining practice, which is devastating Appalachia. Over 1,200 miles (about 1,900 kilometers) of streams have been buried and destroyed and countless mountains and ridgetops have been blown up to extract coal. Website has resource "What's my connection to mountaintop removal?" which allows visitors to enter their zip code to see if their energy is derived from mountaintop removal. See also Mountain Justice, below.
www.ilovemountains.org

MONGABAY.COM
Important resource providing up-to-date news, statistics, photos, and information about tropical forests, including in-depth articles and interviews about the state of the Amazon and other rain forests.
http://rainforests.mongabay.com

MOUNTAIN JUSTICE
A call to action from the people of the Appalachian Mountains, who seek to save their mountains, streams, and forests from coal companies' practice of mountaintop removal. See also iLoveMountains.org, above.
www.mountainjustice.org

PAPER CUTS: RECOVERING THE PAPER LANDSCAPE
1999 Worldwatch Institute report about global paper use, and how it can be reduced by at least 50 percent.
www.worldwatch.org/node/841

PRINCE'S RAINFORESTS PROJECT

Important British organization that works with governments, businesses, NGOs, and individuals to increase global recognition of the contribution of tropical deforestation to climate change and to find ways to make the rain forests worth more alive than dead. Established in 2007 by "HRH The Prince of Wales."

www.princesrainforestsproject.org

RAINFOREST ACTION NETWORK (RAN)

A major rain forest activist organization. Uses hard-hitting market campaigns to align the policies of multinational corporations with widespread public support for environmental protection. Believes that logging ancient forests for copy paper or destroying an endangered ecosystem for a week's worth of oil is not just destructive, but outdated and unnecessary.

www.ran.org

SAVE THE BOREAL FOREST

Imperative group established to save the boreal forests. "The Northern American Boreal forest comprises 25% of the world's last remaining ancient forest," which a handful of US-based multinational corporations, such as Kimberly-Clark and Weyerhaeuser, are systematically destroying. Affiliate websites include www.kleercut.net and www.freegrassy.org. See also the Greenpeace Canada "Save the Boreal Forest" page, www.greenpeace.org/canada/en/campaigns/boreal.

www.savetheboreal.org

***STATE OF THE WORLD'S FORESTS* REPORTS**

Reports by the Food and Agriculture Organization of the United Nations (FAO) on the status of forests, recent major policy and institutional developments, and key issues concerning the world's forests. Published biennially (next edition 2009).

2007 edition: www.fao.org/docrep/009/a0773e/a0773e00.htm

TREE-FREE PAPER

Tree-free paper is increasingly available, as is 100% recycled paper. Here are two sources: Ecopaper (www.ecopaper.com) sells tree-free paper from banana, coffee, and other by-products; GreenLine Paper (www.greenlinepaper.com) sells only tree-free and recycled paper (including office paper). For more tree-free and recycled paper companies, go to www.ecomall.com/biz/paper.htm.

Toxic Pollution

BE SAFE CAMPAIGN FOR PRECAUTIONARY ACTION

Nationwide initiative by the Center for Health, Environment and Justice to build support for the precautionary approach, which looks at how people can prevent harm from environmental hazards, a "better-safe-than-sorry" practice motivated by caution.
www.besafenet.com

BIRDS, BEES AND MANKIND—DESTROYING NATURE BY "ELECTROSMOG"

2008 report by Dr. Ulrich Warnke of the University of Saarland, Germany, which concludes that power lines, Wi-Fi, cell phones, and other sources of "electrosmog" are causing enormous disruption and death to birds and bees around the world.
www.powerwatch.org.uk/news/20080917_warnke_birds_bees.pdf

BREAST CANCER ACTION (BCA)

Group specifically committed to the precautionary principle of public health: do no harm. Encourages the use of environmentally safe alternatives to ways of doing business that we know—or have reason to believe—are harmful. Educates about environmental factors that might cause breast cancer. Produces the online newsletter *BCA Source*.
www.bcaction.org

CENTER FOR ENVIRONMENTAL HEALTH (CEH)

Environmental advocacy group that contends that the powerful companies that make and use industrial chemicals have a dirty secret: synthetic chemicals surround us all, and many are making people sick. Works to eliminate the threat that these substances pose to children, families, and communities.
www.cehca.org

CENTER FOR HEALTH, ENVIRONMENT AND JUSTICE (CHEJ)

New York–based group that helps communities and individuals facing environmental health risks—from leaking landfills and polluted drinking water to incinerators and hazardous-waste sites. Founded by longtime environmental activist Lois Gibbs (who led the movement to relocate families away from the toxic waste dump in Love Canal, New York). CHEJ was instrumental in establishing some of the first national policies critical to protecting community health, like the Superfund program, Right-to-Know, and others.

www.chej.org

CENTERS FOR DISEASE CONTROL AND PREVENTION (CDC): AIR POLLUTION AND RESPIRATORY HEALTH

Group specifically dedicated to leading the CDC's fight against environmental-related respiratory illnesses (including asthma) and studies indoor and outdoor air pollution. Website has articles and documents to educate about various sources of air pollution.

www.cdc.gov/nceh/airpollution

CHEMICAL BODY BURDEN

Website that provides information, case studies, and reports from health professionals, scientists, citizen groups, and environmental organizations concerned about the chemical body burden we all carry and its health effects—known and unknown. Scientists estimate that everyone alive today carries within the body at least seven hundred contaminants, most of which have not been well documented.

www.chemicalbodyburden.org

THE CHEMICAL SENSITIVITY FOUNDATION

Nonprofit corporation that raises public awareness about multiple chemical sensitivity. Website has brochures, books, and DVDs, as well as referrals to environmental medicine physicians.

www.chemicalsensitivityfoundation.org

CITIZENS CONCERNED ABOUT CHLORAMINE

Raises the public's level of awareness about chloramine and its health effects when used as a disinfectant in the water. Pursues and supports scientific studies that will explore the effects of chloramine on humans, animals, and the environment, and supports legislation to protect the quality of water. Website has fact sheets, studies, and reports of hearings where citizens testify about health problems from chloramine, which the group contends is much more harmful than chlorine.

www.chloramine.org

CLEAN WATER ACTION

Environmental action group working to empower people to take action to protect US waters, build healthy communities, and make democracy work for all of us by holding elected officials accountable to the public.

www.cleanwateraction.org

CONTAINER RECYCLING INSTITUTE (CRI)

Nonprofit organization that educates policy makers, government officials, and the general public regarding the social and environmental impacts of the production and disposal of no-deposit, no-return beverage containers and the need for producers to take responsibility for their wasteful packaging. Promotes recovery and recycling.

www.container-recycling.org

DEBRA LYNN DADD

Dadd, a consumer advocate and author of several books about adopting a green lifestyle without toxins, provides several resources on her website, including "Debra's List," which lists hundreds of nontoxic, natural, and earth-wise products; her newsletter and blog on eco-living; and information about recovery from multiple-chemical sensitivity.

www.dld123.com

EM RADIATION RESEARCH TRUST

UK organization that aims to provide the facts about the relationship between electromagnetic radiation—such as power lines, cell phones, and Wi-Fi—and human health. Website has extensive and frequently updated news articles and reports.

www.radiationresearch.org

EMR NETWORK

"Citizens and professionals for the responsible use of electromagnetic radiation." Educates the public about the body of research that exists on the biological effects of low-intensity, nonthermal exposure to non-ionizing radiation, and where this body of knowledge intersects with public health and local land-use regulations.

www.emrnetwork.org

ENVIRONMENTAL HEALTH PERSPECTIVES

Monthly journal of peer-reviewed research and news on the impact of the environment on human health.

www.ehponline.org

ENVIRONMENTAL WORKING GROUP (EWG)

Nonprofit health advocacy organization that uses the power of public information to protect the most vulnerable segments of the human population—children, babies, and infants in the womb—from health problems attributed to toxic contaminants. Advocates replacing harmful federal policies, including government subsidies that damage the environment and natural resources, with policies that invest in conservation and sustainable development.

www.ewg.org

FLUORIDE ACTION NETWORK

An international coalition seeking to broaden public awareness about the toxicity of fluoride compounds and the health impacts of current fluoride exposures. To read the "Professionals' Statement Calling for an End to Water Fluoridation" (2007), signed by over 2,000 professionals including 267 doctors, go to www.fluoridealert.org/prof-statement.pdf. To read a 500-page review of fluoride's toxicity, *Fluoride in Drinking Water: A Scientific Review of EPA's Standards*, go to www.nap.edu/catalog.php?record_id=11571.

www.fluoridealert.org

HEALTHY CHILD HEALTHY WORLD

Pediatric preventive group that helps parents, educators, health professionals, and the general public take action to create healthy environments and embrace green, nontoxic practices. Exists because more than 125 million Americans, especially children, now face a historically unprecedented rise in illnesses such as cancer, autism, asthma, birth defects, ADD/ADHD, and learning and developmental disabilities, many of which are likely caused by environmental hazards and household chemicals.

www.healthychild.org

INTERNATIONAL DARK-SKY ASSOCIATION (IDA)

Nonprofit organization that focuses on preserving the nighttime environment; stopping the adverse effects of light pollution on humans, wildlife, and ecosystems; and reducing the energy waste from excess or inefficient lighting. Two relevant articles are the November 2008 *National Geographic* cover story about light pollution, "Our Vanishing Night" (http://ngm

.nationalgeographic.com/2008/11/light-pollution/klinkenborg-text); and the February 20, 2008 *Washington Post* story "Lights at Night Are Linked to Breast Cancer" (www.washingtonpost.com/wp-dyn/content/article/2008/02/19/AR2008021902398.html?nav=rss_health).

www.darksky.org

MOBILE PHONES AND BRAIN TUMORS—A PUBLIC HEALTH CONCERN

Research paper (69 pages) on the use of cell phones and incidence of brain tumors, by Vini Khurana, MD, neurosurgeon at Canberra Hospital, Australia. Dr. Khurana (www.brain-surgery.net.au/c_a.html) reviewed over a hundred studies on the effects of cell phones. See also *Tumors and Cell Phone Use: What the Science Says*, below.

www.brain-surgery.us/mobph.pdf

PESTICIDE ACTION NETWORK OF NORTH AMERICA

With offices on all major continents, works to replace the use of hazardous pesticides with ecologically sound and socially just alternatives. Links local and international consumer, labor, health, environment, and agricultural groups into an international citizens' action network that challenges the global proliferation of pesticides, defends basic rights to health and environmental quality, and works to ensure the transition to a just and viable soceity.

www.panna.org

SCIENCE & ENVIRONMENTAL HEALTH NETWORK

Leading proponent in the United States of the precautionary principle as a new basis for environmental and public health policy. Works with issue-driven organizations, national environmental health coalitions, governments, and NGOs to implement precautionary policies at local and state levels, and to apply science in effective ways to protect and restore public and ecosystem health.

www.sehn.org

SILENT SPRING INSTITUTE

Institute, inspired by Rachel Carson, that builds on a unique partnership of scientists, physicians, public health advocates, and community activists to identify and break the links between the environment and women's health, especially breast cancer. Recent reports have focused on household and personal-care products, flame retardants, pharmaceuticals, and hormones.

www.silentspring.org

TUMORS AND CELL PHONE USE: WHAT THE SCIENCE SAYS

Study (84 pages) by Ronald B. Herberman, MD, director of the University of Pittsburgh Cancer Institute, of nine thousand cell phone users, which finds that cell phones could cause cancer, and that children are especially at risk. According to Dr. Herberman (www.upmccancercenters.com/about/bio-herberman.html), "I would really hate to stand by and wait for an epidemic of brain tumors to occur before something is done about it." To read the article "Pittsburgh Cancer Center Warns of Cell Phone Risks" go to http://abcnews.go.com/Health/wireStory?id=5439074. See also *Mobile Phones and Brain Tumors—A Public Health Concern*, above.

http://domesticpolicy.oversight.house.gov/documents/20080925142803.pdf

U.S. PUBLIC INTEREST RESEARCH GROUP (PIRG): TOXIC-FREE COMMUNITIES

Ralph Nader's organization working to create and implement safer, healthier alternatives to industrial pollution. In more than seventeen states, its advocates and organizers make the case for key policies needed to protect public health and the environment from toxic chemicals.

www.uspirg.org/issues/healthy-communities

WOMEN'S ENVIRONMENTAL NETWORK

UK organization that increases awareness of women's perspectives on environmental issues and influences decision making to achieve environmental justice for women.

www.wen.org.uk

Nuclear Waste / Military Pollution

BEYOND NUCLEAR

Washington DC–based group dedicated to educating the public about the connection between nuclear power and nuclear weapons, and the need to abandon both to safeguard the future. Teaches that nuclear energy is not a solution to global warming, and illuminates the dangers of nuclear waste.

www.beyondnuclear.org

MILITARY TOXICS PROJECT

National nonprofit network of neighborhood, veterans', indigenous, peace, environmental, and other grassroots organizations representing people affected by military contamination and pollution. Works with community leaders to help them act and speak for themselves. Focuses on ensuring cleanup of military pollution, limiting the transport of hazardous materials, and advancing preventive solutions to the toxic and radioactive pollution caused by military activities.

www.stopmilitarytoxics.org

NUCLEAR INFORMATION AND RESOURCE SERVICE (NIRS)

National information and networking center for citizens and environmental activists concerned about nuclear power, radioactive waste, radiation, and sustainable energy. Initiates large-scale organizing and public education campaigns on specific issues, such as preventing construction of new reactors, transportation of radioactive waste, deregulation of radioactive materials, and more.

www.nirs.org

RADIATED VETERANS OF AMERICA (RVA)

Veterans service organization that, through public education and endorsement of legislation, aims to aid past, current, and future members of the US Armed Forces who, beginning with the Manhattan Project in 1945, were or will be exposed to ionizing radiation from weapons tests, storage, accidents, and deployment. Includes nonweapons use of nuclear energy for medical, propulsion, and experimental purposes. Members include veterans exposed to depleted uranium (DU) in the Gulf, Kosovo, Afghanistan, and Iraq wars, and those exposed to Agent Orange.

www.radvets.org

WORLD INFORMATION SERVICE ON ENERGY (WISE) URANIUM PROJECT

Information and networking center for citizens and environmental organizations concerned about the health and environmental impacts of nuclear energy, radioactive waste, radiation, and related issues. Collaborates with the US-based group Nuclear Information and Resource Service (NIRS), including copublishing twenty times a year the *WISE/NIRS Nuclear Monitor*, an international newsletter serving the worldwide movement against nuclear power.

www.wise-uranium.org

Water Filters

To avoid using plastic bottles for drinking water, a good home water filter is recommended. Here are two companies that sell high-quality water filters.

THE CUTTING EDGE
Offers an extensive line of water filters and distillers.
www.cutcat.com

RENEWED HEALTH
Sells the Precision line of water distillers.
www.renewedhealth.com

Species Extinction

CENTER FOR BIODIVERSITY
Organization that believes that the welfare of human beings is deeply linked to the existence of a vast diversity of wild animals and plants. Works to secure a future for all species, great and small, hovering on the brink of extinction. Focuses on protecting the lands, waters, and climate that species need to survive, with a goal that future generations will inherit a world in which the wild is still alive.
www.biologicaldiversity.org

DEFENDERS OF WILDLIFE
Goup specifically dedicated to the preservation of all wild animals and native plants in their natural communities, because, as President Rodger Schlickeisen puts it, "nearly everywhere, natural habitats—home to all kinds of creatures—are being destroyed to make way for new shopping malls, roads and housing developments." Champions the Endangered Species Act, preventing oil drilling in public lands, and stopping the exploitation and killing of animals for pets, food, or souvenirs. Offers *Defenders* magazine to members and free *Wildlife* e-news bulletin to everyone.
www.defenders.org

INTERNATIONAL FUND FOR ANIMAL WELFARE
Engages communities, government leaders, and like-minded organizations around the world to achieve lasting solutions for pressing animal

welfare and conservation challenges—solutions that will benefit both animals and people. Rejects the notion that the interests of humans and animals are separate. Works to protect seals, dolphins, whales, bears, elephants, and primates and other animals.
www.ifaw.org

INTERNATIONAL UNION FOR THE CONSERVATION OF NATURE (IUCN): SPECIES PROGRAMME

Important organization that produces, maintains, and manages the annual IUCN "Red List of Threatened Species." Implements global species conservation initiatives, including Red List Biodiversity Assessment projects to assess the status of species for the Red List. Calls itself "the world's oldest and largest global environmental network—a democratic membership union with more than 1,000 government and NGO member organizations."
www.iucn.org/about/work/programmes/species

NATIONAL INVASIVE SPECIES INFORMATION CENTER

Established by the USDA National Agricultural Library to provide information on invasive species that harm and displace native animals and plants. Two other websites with invasive-species information are www.issg.org (The Global Invasive Species Database) and www.invasive.org (The Nature Conservancy's archived invasive-species page).
www.invasivespeciesinfo.gov

WORLD WILDLIFE FUND (WWF)

International environmental organization whose mission is the conservation of nature. Publishes the biennial *Living Planet Report.* Works in a hundred countries. With a foundation in science, takes action to ensure the delivery of innovative solutions that meet the needs of both people and nature.
www.wwf.org

Population Issues

POPULATION ACTION INTERNATIONAL

International group that works to ensure that every person has the right and access to sexual and reproductive health, so that humanity and the natural environment can exist in balance and fewer people live in poverty. Fosters the development of US and international policy on urgent population and reproductive health issues through an integrated pro-

gram of research, advocacy, and communication. Seeks to make clear the links between population, reproductive health, the environment, and development.
www.populationaction.org

POPULATION CONNECTION
National grassroots organization that educates young people and advocates progressive action to stabilize world population at a level that can be sustained by Earth's resources. Formerly Zero Population Growth.
www.populationconnection.org

Food Issues / World Hunger

BREAD FOR THE WORLD
A collective Christian voice urging decision makers to end hunger at home and abroad. By changing policies, programs, and conditions that allow hunger and poverty to persist, provides help and opportunity far beyond the communities in which we live.
www.bread.org

CENTER FOR FOOD SAFETY (CFS)
Nonprofit public interest and environmental advocacy organization that works to protect human health and the environment by curbing the proliferation of harmful food production technologies and promoting organic and other forms of sustainable agriculture. Website has much information on genetically engineered food, food and global warming, and food irradiation, plus action items and news stories.
www.centerforfoodsafety.org

FOOD AND AGRICULTURE ORGANIZATION OF THE UNITED NATIONS (FAO)
Arm of the United Nations that leads international efforts to defeat hunger; modernize and improve agriculture in rural areas, forestry, and fishery practices; and ensure good nutrition for all. 2006 report *Livestock's Long Shadow* (www.fao.org/docrep/010/a0701e/a0701e00.htm) describes how livestock are responsible for 18 percent of greenhouse gas emissions. 2008 report *Food Energy and Climate: A New Equation* (ftp://ftp.fao.org/docrep/fao/011/i0330e/i0330e00.pdf) explains how the poor will be disproportionately affected by global warming and the growing of food for biofuels.
www.fao.org

FOOD & WATER WATCH

Organization that challenges harmful food production technologies and promotes sustainable alternatives. Opposes growth hormones in milk, factory farming, and the takeover of small farms and the global oceans by multinational corporations.

www.foodandwaterwatch.org

GREENPEACE: GENETIC ENGINEERING

Information about the dangers of genetically engineered food.

www.greenpeace.org/international/campaigns/genetic-engineering

THE HUNGER PROJECT

Global nonprofit, strategic organization committed to the sustainable end of world hunger. In Africa, Asia, and Latin America, seeks to end hunger and poverty by empowering people to lead lives of self-reliance, meet their own basic needs, and build better futures for their children.

www.thp.org

INTERNATIONAL SOCIETY FOR ECOLOGY & CULTURE (ISEC)

Nonprofit organization concerned with the protection of both biological and cultural diversity. Has established an Ancient Futures Network to bring together groups and individuals from every corner of the world that are struggling to maintain their cultural integrity in the face of economic globalization.

www.isec.org.uk

NATIONAL GARDENING ASSOCIATION: KIDSGARDENING

Gardening group "helping young minds grow" with teacher and parent resources, projects, *Kids Garden News* (e-mail update), and online store.

www.kidsgardening.com

ORGANIC CONSUMERS ASSOCIATION

Important group that campaigns for health, justice, and sustainability. Deals with food safety, industrial agriculture, genetic engineering, children's health, corporate accountability, fair trade, environmental sustainability, and other key topics. Promotes the interests of the United States' estimated fifty million organic and socially responsible consumers.

www.organicconsumers.org

RESULTS

Grassroots advocacy organization committed to creating the political will to end hunger and the worst aspects of poverty, by lobbying elected officials for effective solutions and key policies that affect hunger and poverty.

www.results.org

SEED SAVERS EXCHANGE

Nonprofit organization that saves and shares heirloom seeds, forming a living legacy that can be passed down through generations to nurture our diverse, fragile, genetic and cultural garden heritage. Since 1975, members have passed on approximately one million samples of rare seeds to other gardeners. Visitor center in Dacorah, Iowa.

www.seedsavers.org

SMALL PLANET INSTITUTE

Organization founded to support grassroots democracy movements worldwide that address the causes of hunger and poverty. Works to increase participation in democracy. Focuses particularly on the issue of ending world hunger. Cofounder Frances Moore Lappé wrote *Diet for a Small Planet.*

www.smallplanet.org

WORLD FOOD PROGRAMME

The United Nations frontline agency in the fight against global hunger. It is the world's largest humanitarian organization. Our food assistance reaches an average of 100 million people in 80 countries every year. Almost 12,000 people work for the organization, most of them in remote areas, directly serving the hungry poor.

www.wfp.org

Soil Erosion

GEOMORPHOLOGICAL RESEARCH GROUP

Research group that operates under supervision of Dr. David R. Montgomery, author of *Dirt: The Erosion of Civilization. Nature* wrote that *Dirt* "insightfully chronicles the rise of agricultural technology and concomitant fall of soil depth just about everywhere in the world . . . We are currently losing soil 20 times faster, on average, than it is being replaced through the natural process. To meet the demands for food and, more recently, energy,

we need Montgomery's scholarly, historical perspective, as well as the ability to project current trends of land management to future scenarios."
http://duff.ess.washington.edu/grg

Consumerism

THE COMPACT

Movement advocating "a 12-month flight from the consumer grid" and not buying anything new. Goals are (1) "to go beyond recycling in trying to counteract the negative global environmental and socioeconomic impacts of disposable consumer culture and to support local businesses, farms, etc."; (2) "to reduce clutter and waste in our homes"; (3) "to simplify our lives." Written about in the article "Out of the Retail Rat Race: Consumer Group Doesn't Buy Notion That New Is Better" (www.sfgate.com/cgi-bin/article.cgi?f=/c/a/2006/02/13/BAGH3H7DH71.DTL&hw=freecycling&sn=004&sc=409).
http://sfcompact.blogspot.com (links to many other chapters around the world)
http://groups.yahoo.com/group/thecompact

FREECYCLE

Nonprofit grassroots movement that, by providing a network in which people can give (and get) stuff for free in their own towns, promotes waste reduction to save the landscape from being taken over by landfills. Made up of 4,642 groups with nearly 6.2 million members across the globe.
www.freecycle.org

REVEREND BILL: THE CHURCH OF STOP SHOPPING

Advocacy group led by comic preacher. Opposes chain stores and malls that hurt neighborhoods.
www.revbilly.com

THE ULS REPORT

A bimonthly newsletter created to help people Use Less Stuff by conserving resources and reducing waste. Created a Sustainable Products Certification Program that allows companies to feature the ULS logo on their packages if they can prove significant reductions in material and energy consumption.
http://use-less-stuff.com

Corporations

CORPORATE ACCOUNTABILITY INTERNATIONAL

Group that demands corporate responsibility. Has been waging winning campaigns to challenge corporate abuse for more than thirty years. Demands direct corporate accountability to public interests. Current campaigns focus on water access, junk food, and tobacco marketing and associated companies.
www.stopcorporateabuse.org/cms

CORPORATE WATCH

The rights of corporations—disguised as "encouraging foreign investment," "promoting free trade," and "protecting the national interest"—now take precedence over human rights, community interests, and the interests of the planet itself. Corporate Watch is a research group supporting campaigns that are increasingly successfully forcing corporations to back down from environmentally destructive or socially divisive projects.
www.corporatewatch.org

THE CORPORATION

Award-winning documentary film focused on the nature and spectacular rise of the dominant institution of our time. Based on the book *The Corporation: The Pathological Pursuit of Profit and Power* by Joel Bakan. DVD includes 145-minute theatrical release and eight hours of extras.
www.thecorporation.com

CORPORATIONS AND PROPAGANDA

TUC Radio's downloadable one-hour program, in two parts, based on Australian Alex Carey's research about how corporations captured and continue to dominate the democratic process in the United States.
www.radio4all.net/index.php/program/4041

INTERNATIONAL FORUM ON GLOBALIZATION

Research and educational institution that provides analyses on the cultural, social, political, and environmental impacts of economic globalization, with a shared concern that the world's corporate and political leadership has taken over and restructured global politics and economics. Closely linked to social justice and environmental movements.
www.ifg.org

MULTINATIONAL MONITOR

Bimonthly online news magazine focused on the power of transnational companies. Publishes "10 Worst Corporations" list each year. For the 2008 list, go to www.multinationalmonitor.org/mm2008/112008/weissman.html.

www.multinationalmonitor.org

PUBLIC CITIZEN: GLOBAL TRADE WATCH

Part of Ralph Nader's public-interest, consumer advocacy organization Public Citizen that promotes democracy by challenging corporate globalization, arguing that the current globalization model is neither a random inevitability nor "free trade."

www.citizen.org/trade

SPRAWL-BUSTERS

Helps local community coalitions design and implement successful campaigns against megastores and other undisirable large-scale developments. Their slogan is"Home Town America Fights Back!" Founder Al Norman has written *Slam-Dunking Wal-Mart: How You Can Stop Superstore Sprawl in Your Hometown* and *The Case Against Wal-Mart.*

www.sprawl-busters.com

THE 10 WORST CORPORATIONS

Multinational Monitor's annual list of the most destructive companies. The 2008 list (at www.multinationalmonitor.org/mm2008/112008/weissman.html) includes AIG, Chevron, Dole, and General Electric. List appears in each year's November/December issue.

WAL-MART WATCH

Nationwide public education campaign in the United States to challenge the world's largest retailer, Wal-Mart, to become a better employer, neighbor, and corporate citizen. Bridges the gap between ordinary citizens and community organizations concerned about Wal-Mart's unchecked growth and negative impact on our soceity. Challenges Wal-Mart to embrace its moral responsibility as the nation's biggest and most important corporation.

www.walmartwatch.com

Media Accountability

CENTER FOR PUBLIC INTEGRITY: MEDIA TRACKER

Media-monitoring group that identifies who owns the media where you live. Visitors can enter their zip code or a city and state, and it combs a database of more than five million records from government sources, corporate documents, and original research.

http://projects.publicintegrity.org/telecom

FAIRNESS & ACCURACY IN REPORTING (FAIR)

National media watch group offering well-documented criticism of media bias and censorship. Exposes neglected news stories. Believes that structural reform is ultimately needed to break up the dominant media conglomerates, establish independent public broadcasting, and promote strong nonprofit sources of information.

www.fair.org

INSTITUTE FOR PUBLIC ACCURACY (IPA)

Organization that seeks to broaden public discourse. News releases offer well-documented analysis of current events and underlying issues, and promote the inclusion of perspectives that widen the bounds of media discussion and enhance democratic debate. Founded by journalist Norman Solomon.

www.accuracy.org

MEDIACHANNEL

Global nonprofit online network that is concerned with the political, cultural, and social impacts of the media, large and small. Provides information and diverse perspectives to inspire debate, collaboration, action, and citizen engagement.

www.mediachannel.org

PROJECT CENSORED

Media research program that conducts research on important national news stories that are underreported, ignored, misrepresented, or censored by the US corporate media. Each September, Project Censored publishes a ranking of the twenty-five most censored nationally important news stories.

www.projectcensored.org

Environmental News and Analysis

PRINT AND ONLINE RESOURCES

ALTERNET: ENVIRONMENT

Environmental news page of *AlterNet*, a news magazine and online community that creates original journalism and amplifies the best of hundreds of other independent media sources. *AlterNet*'s aim is to inspire action and advocacy on the environment and other issues.

www.alternet.org/environment

COMMON DREAMS

Daily compendium of news and feature stories from world newspapers, often featuring environmental news and commentary.

www.commondreams.org

DOT EARTH

New York Times reporter Andrew C. Revkin's green blog.

http://dotearth.blogs.nytimes.com

E MAGAZINE

Monthly environmental magazine (print and online).

www.emagazine.com

EARTH ISLAND JOURNAL

Quarterly environmental news magazine (print and online).

www.earthisland.org/journal

ENVIRONMENTAL HEALTH NEWS

Website that advances the public's understanding of envirnomental health issues by providing access to worldwide news about a variety of subjects related to the health of humans, wildlife, and ecosystems.

www.environmentalhealthnews.org

ENVIRONMENTAL HEALTH PERSPECTIVES

Monthly journal of peer-reviewed research and news on the impact of the environment on human health.

www.ehponline.org

ENVIRONMENTAL NEWS NETWORK (ENN)
Exhaustive coverage of environmental news from around the world.
www.enn.com

GRIST
Environmental news and commentary. Attempts to tell untold stories, spotlight trends, and engage readers with independent perspective.
http://gristmill.grist.org

***GUARDIAN*: ENVIRONMENT**
Environmental news and commentary from the *Guardian*, a British newspaper.
www.guardian.co.uk/environment

***HUFFINGTON POST*: GREEN**
Environmental news page from Arianna Huffington's website of progressive commentary.
www.huffingtonpost.com/tag/environment

***INDEPENDENT*: ENVIRONMENT**
Environmental news and commentary from the *Independent*, a British newspaper.
www.independent.co.uk/environment

ONEARTH
Natural Resources Defense Council's monthly magazine on environmental affairs, particularly those relating to policies.
www.onearth.org

REUTERS: ENVIRONMENT
Environmental stories from Reuters news service.
www.reuters.com/news/environment

SOCIETY OF ENVIRONMENTAL JOURNALISTS (SEJ)
Journalistic group that advances public understanding of environmental issues by improving the quality, accuracy, and visibility of environmental reporting. Provides critical support to journalists of all media in their efforts to cover complex issues of the environment responsibly. Website has many investigative news stories each day.
www.sej.org

TREEHUGGER

Website dedicated to driving sustainibility mainstream with green news, solutions, and product information. Features a blog, weekly and daily newsletters, weekly video segments, weekly radio show, and a user-generated blog, Hugg.

www.treehugger.com

RADIO

DEMOCRACY NOW!

Weekday, two-hour public-affairs radio and television program hosted by Amy Goodman and Juan Gonzalez. Some programs feature environmental topics.

www.democracynow.org

EARTHBEAT RADIO

Weekly, one-hour environmental talk radio show.

www.earthbeatradio.org

ECOSHOCK

The Internet's "largest green audio download site," and its only twenty-four-hour environmental radio station.

www.ecoshock.org

FLASHPOINTS

Weekday, one-hour radio show hosted by Dennis Bernstein. Covers political, economic, environmental, and other topics.

www.flashpoints.net

GLOBAL PUBLIC MEDIA

Public information service that presents radio programs on environmental and other topics.

www.globalpublicmedia.com

IF YOU LOVE THIS PLANET

Dr. Helen Caldicott's weekly, one-hour radio program about environmental, nuclear, and other public health issues.

www.ifyoulovethisplanet.org

LIVING ON EARTH

National Public Radio's (NPR) weekly, half-hour magazine-style environmental radio program.

www.loe.org

SIERRA CLUB RADIO

Weekly, half-hour radio program.

http://sierraclub.typepad.com/sierra_club_radio

TIME OF USEFUL CONSCIOUSNESS

TUC Radio's weekly, half-hour radio program devoted to environmental, economic, political, and nuclear issues that pays attention to the influence of transnational corporations.

www.tucradio.org

TELEVISION

PLANET GREEN

Twenty-four-hour television and online environmental network, offering more than 250 hours of original green lifestyle programming per month.

www.planetgreen.discovery.com

SUNDANCE CHANNEL: THE GREEN

Television programming devoted entirely to the environment, presented by Robert Redford.

www.sundancechannel.com/thegreen

Educating Children

CENTER FOR ECOLITERACY

Center dedicated to education for sustainable living for K–12 educators, parents, and other members of the school community, based on four guiding principles: (1) "nature is our teacher," (2) "sustainability is a community practice," (3) "the real world is the optimal learning environment," and (4) "sustainable living is rooted in a deep knowledge of place."

www.ecoliteracy.org

THE VIDEO PROJECT

Educational media and documentary programming on environmental issues for schools, libraries, and educators worldwide, with programs from over two humdred independent filmmakers worldwide. Exclusive or primary distributor for most of the programs in their collection. Distributes over ten thousand programs every year.

www.videoproject.com

Notes

Introduction

1. Tara Bradley-Steck, "Expert: Americans Cause Much Damage to the Planet," *Philadelphia Inquirer*, April 6, 1990.
2. "Internet Usage Statistics—The Big Picture," Internet World Stats, www.internetworldstats.com/stats.htm (accessed April 6, 2007); "World Population Trends," US Census Bureau, www.census.gov/ipc/www/idb/worldpopinfo.html (accessed April 6, 2007); Andrew Buncombe, "US Population Hits 300 Million, but Is It Sustainable?" *Independent* (UK), October 11, 2006, www.stwr.org/united-states-of-america/-us-population-hits-300-million-but-is-it-sustainable.html (accessed April 23, 2007).

1. Ozone Depletion

1. *Global Environment Outlook: Environment for Development, GEO 4* (Nairobi, Kenya: United Nations Environment Programme, 2007).
2. Geoffrey Lean, *Action on Ozone* (Nairobi, Kenya: Information and Public Affairs UNEP, 1989).
3. Michael D. Lemonick, "Deadly Danger in a Spray Can," *Time*, January 2, 1989, 42.
4. Lean, *Action on Ozone*.
5. *Scientific Assessment of Ozone Depletion: 2006* (Geneva, Switzerland: World Meteorological Association, 2007), Q.28–Q.29, http://esrl.noaa.gov/csd/assessments/2006/report.html.

6. Lean, *Action on Ozone.*
7. "2007 Skin Cancer Facts," Skin Cancer Foundation, www.skincancer.org/2007-Skin-Cancer-Facts.html (accessed June 3, 2007).
8. *A Greenpeace Australian Strategy to Protect the Ozone Layer* (Balmain, Australia: Greenpeace, 1990).
9. Julian Cribb, "CFC Limits Will Take 14 Years to Reverse UV Trend," *Australian*, September 10, 1991.
10. Thomas E. Graedel and Paul I. Crutzen, "The Changing Atmosphere," Scientific American, September 1989, 58–68.
11. *Greenpeace Australian Strategy.*
12. *Global Environment Outlook.*
13. "Space Shuttle," in *Internet Encyclopedia of Science*, www.daviddarling.info/encyclopedia/S/Space_Shuttle.html.
14. Sharon Ebner, "Solid Fuel Critics Say Ozone Is in Danger," *Sun Herald* (Mississippi), March 17, 1990; Helen Caldicott and Craig Eisendrath, *War in Heaven: The Arms Race in Outer Space* (New York: New Press, 2007).
15. Lean, *Action on Ozone.*
16. Ibid.
17. Margaret Harris, "Skin Cancer Out of Control," *Sydney Morning Herald*, September 12, 1990.
18. "Melanoma Statistics," Melanoma Center, www.melanomacenter.org/basics/statistics.html (accessed June 7, 2007).
19. Susan M. Swetter, "Malignant Melanoma," *eMedicine*, January 23, 2008, http://emedicine.medscape.com/article/1100753-overview.
20. Ibid.
21. Wendy Hansen, "Melanoma Cases Surge among Young Women," *Los Angeles Times*, July 11, 2008.
22. Matthew Blondin, "Eye Diseases Associated with Ultraviolet Light Exposure," PowerPoint presentation, slide 16, May 2, 2007, www.drblondin.com (accessed October 25, 2007).
23. "Health and Environment. International Day for the Protection of the Ozone Layer" (September 16, 2006), United Nations Environment

Programme, www.uneptie.org/Ozonaction/information/education -packlaunch.htm.

24. Lean, *Action on Ozone.*
25. Ibid.
26. Ibid.
27. "US Launch Vehicles," in *The Global Ecology Handbook: What You Can Do about the Environmental Crisis*, ed. Walter H. Corson (Boston: Beacon Press, 1990), 230.
28. Ibid.
29. Stephen Leahy, "Why Is the Ozone Hole Growing?" (September 14, 2005), Inter Press Service, www.commondreams.org/cgi-bin/print.cgi?file=/headlines05/0914-02.htm.
30. "27th Multilateral Fund Meeting Yields Mixed Results" (September 1999), Friends of the Earth, www.foe.org.
31. "CSI 006—Production and Consumption of Ozone Depleting Substances" (assessment draft created 2007), European Environment Agency, http://themes.eea.europa.eu/IMS/ISpecs/ISpecification20041007131537/IAssessment1229857558178/view_content.
32. *Global Environment Outlook.*
33. Helen Threw, "Thinning Threat to Ozone Layer," *Daily Telegraph*, September 22, 1990.
34. George Tyler Miller, *Living in the Environment: Principles, Connections, and Solutions* (Pacific Grove, CA: Thomson Brooks/Cole, 2005), 561.
35. David de Jager, Martin Manning, and Lambert Kuijpers, "Safeguarding the Ozone Layer and the Global Climate System: Issues Related to Hydrofluorocarbons and Perfluorocarbons—Technical Summary," IPCC/TEAP Special Report, www.mnp.nl/ipcc/pages_media/SROC-final/SROC_TS.pdf.
36. Jen Graham, "Global Warming: Smackdown on the Skeptics," *Stanford Daily*, April 19, 2006.
37. Helen Caldicott, *Nuclear Power Is Not the Answer* (New York: New Press, 2006).
38. Margaret Roosevelt, "A Climate Threat from Flat TVs, Microchips," *Los Angeles Times*, July 8, 2008.

2. Global Warming

1. Jim Hansen, "Threat to the Planet," *New York Review of Books*, July 13, 2006.
2. Robin Clarke, *The Greenhouse Gases* (Nairobi, Kenya: United Nations Environment Programme, 1987).
3. "Global Warming: 'Tragedy of the Commons' Revisited" (February 13, 2005), Agence France-Presse, www.commondreams.org/headlines05/0213-06.htm.
4. Elizabeth Kolbert, "The Darkening Sea," *The New Yorker*, November 20, 2006.
5. Lester R. Brown et al., *State of the World, 1990: A Worldwatch Institute Report on Progress towards a Sustainable Society* (New York: Norton, 1990), 18.
6. "The Global Carbon Cycle," *UNESCO/Scope Policy Briefs* 2 (2006).
7. David Spratt and Philip Sutton, *Climate Code Red: The Case for a Sustainable Emergency* (Fitzroy, Australia: Friends of the Earth, 2008), www.foe.org.au/resources/publications/climate-justice/climatecodered.pdf.
8. James Hansen, Makiko Sato, Pushker Kharecha, David Beerling, Valarie Masson-Delmotte, Mark Pagani, Maureen Raymo, Dana Royer, and James C. Zachos, "Target Atmospheric CO_2: Where Should Humanity Aim?" unpublished manuscript, April 2008.
9. Spratt and Sutton, *Climate Code Red*.
10. Elisabeth Rosenthal, "China Increases Lead as Biggest Carbon Dioxide Emitter," *New York Times*, June 14, 2008.
11. Bea Vongdouangchanh, "Carbon Capture and Storage 'Being Oversold as a Panacea,'" *Hill Times*, April 13, 2009.
12. Mark Diesendorf, "Can Geosequestration Save the Coal Industry?" in *Transforming Power: Energy, Environment, and Society in Conflict* (Energy and Environmental Policy Series, vol. 9), eds. John Byrne, Noah Toly, and Leigh Glover (New Brunswick, NJ: Transaction Press, 2006).
13. Hansen et al., "Target Atmospheric CO_2."
14. Spratt and Sutton, *Climate Code Red*.
15. Ibid.

16. "Climate Change Hitting Arctic Faster, Harder—Polar Bears May Be at Even Greater Risk" (April 23, 2008), WWF-Canada, http://wwf.ca/newsroom/?1361.

17. Zachary Coile, "Rush to Arctic as Warming Opens Oil Deposits," *San Francisco Chronicle*, August 12, 2008.

18. Ibid.

19. J. E. Hansen, "Scientific Reticence and Sea Level Rise," *Environmental Research Letters* 2 (2007): 024002.

20. Nicholas Stern, "Stern Review: The Economics of Climate Change—Executive Summary" (October 2006), www.hm-treasury.gov.uk/d/Executive_Summary.pdf.

21. Walter H. Corson, ed., *The Global Ecology Handbook: What You Can Do about the Environmental Crisis* (Boston: Beacon Press, 1990), 232.

22. Fred Pearce, "Cities May Be Abandoned as Salt Water Invades," *New Scientist*, April 16, 2006.

23. Ibid.

24. Spratt and Sutton, *Climate Code Red*.

25. Ibid.

26. Juliet Eilperin, "Report Says Severe Weather to Increase as Earth Warms," *Washington Post*, June 21, 2008.

27. Phillip Coorey and Stephanie Peatling, "The Garnaut Ultimatum; Adapt or Perish," *Sydney Morning Herald*, July 5–6, 2008.

28. Mark Davis, "Heatwave Every Year, Scientists Predict," *Sydney Morning Herald*, July 7, 2008.

29. Mark Townsend and Paul Harris, "Now the Pentagon Tells Bush: Climate Change Will Destroy Us. Secret Report Warns of Rioting and Nuclear War; Threat to the World Is Greater than Terrorism," *Observer* (UK), February 22, 2004.

30. Don Comis, "Monitoring Methane—Agriculture Research Service," *Journal of Agricultural Research*, June 1995.

31. Mike Seccombe, "An Ill Wind That Only Does Cows Good," *Sydney Morning Herald*, July 18, 1989.

32. Jim Motavalli, "The Meat of the Matter; Our Livestock Industry Creates More Greenhouse Gas than Transportation," *E Magazine*, July/August 2008.

33. "Global Warming," in *Encarta* (Microsoft), http://encarta.msn.com/encyclopedia_761567022/Global_Warming.html.

34. Michael McCarthy and Clare Kenny, "Environmental Chiefs Join Forces to Fight Growth in Air Travel," *Independent* (UK), June 27, 2006.

35. "Another Culprit of the Greenhouse Effect: Jet Aircraft," *Sydney Morning Herald*, August 27, 1991.

36. Spratt and Sutton, *Climate Code Red*.

37. Elisabeth Rosenthal, "Air Travel and Carbon on Increase in Europe," *New York Times*, June 22, 2008.

38. Gary Stoller, "Concern Grows over Pollution from Jets," *USA Today*, December 19, 2006.

39. Ian Sample, "Not Just Warmer: It's the Hottest for 2,000 Years; Widest Study Yet Backs Fears over Carbon Dioxide," *Guardian* (UK), September 1, 2003.

40. Clarke, *Greenhouse Gases*; Stephen Schneider, "The Changing Climate," *Scientific American*, September 1989; Michael Lemonick, "Feeling the Heat," *Time*, January 2, 1989.

41. *Climate Change 2007: Impacts, Adaptation and Vulnerability: Contribution of Working Group II to the Fourth Assessment Report of the Intergovernmental Panel on Climate Change* (Cambridge: Cambridge University Press, 2007).

42. Schneider, "Changing Climate."

43. Lemonick, "Feeling the Heat."

44. Ibid.

45. Stern, "Stern Review."

46. J. Borger, "Climate Change Disaster Is upon Us, Warns UN," *Guardian* (UK), October 5, 2007.

47. James Murray, "Stern Admits Report 'Badly Underestimated' Climate Change Risks," *Business Green*, April 17, 2008.

48. *Blown Away: How Global Warming Is Eroding the Availability of Insurance Coverage in America's Coastal States* (Environmental Defense, 2007), www.edf.org/documents/7301_BlownAway_insurancereport.pdf.

49. Stephen Leahy, "The Big Melt Coming Faster Than Expected," *Interpress Service*, April 5, 2006.

50. Jitendra Joshi, "US Suffers World's First Climate Change Exodus: Study," *Agence France-Presse*, August 17, 2006.
51. "Burma Still Reeling in Wake of Savage Blow," *Washington Post*, July 7, 2008.
52. Steve Connor, "Global Warming Fastest for 20,000 Years—and It Is Mankind's Fault," *Independent* (UK), May 4, 2006.
53. Emma Ross, "Study Shows Antarctic Glaciers Shrinking," Associated Press, April 22, 2005.
54. Roland Buerk, "Flooded Future Looms for Bangladesh" (December 7, 2004), BBC News, http://news.bbc.co.uk/1/hi/sci/tech/4056755.stm.
55. Deborah Smith, "Sea Warmth Rise Worse Than Was Thought," *Sydney Morning Herald*, June 19, 2008.
56. Lemonick, "Feeling the Heat."
57. Steve Connor, "Climate Change Is Killing the Oceans' Microscopic Lungs," *Independent* (UK), December 7, 2006.
58. Schneider, "Changing Climate."
59. Clarke, *Greenhouse Gases*.
60. Corson, *Global Ecology Handbook*, 233.
61. Clarke, *Greenhouse Gases*; Schneider, "Changing Climate."
62. "Iowa's Disasters," *New York Times*, June 14, 2008.
63. "Rain a Fresh Disaster for South China," *Australian*, June 17, 2008, www.theaustralian.news.com.au/story/0,,23873697-25837,00.html.
64. Schneider, "Changing Climate."
65. Jane Kay, "If You Thought Last Week Was Hot … /Higher Temperatures, Rising Ocean, Loss of Snowpack Forecast for State," *San Francisco Chronicle*, August 1, 2006.
66. "First Half of 2006 Is Warmest on Record," Associated Press, July 14, 2006.
67. "Where There's Smoke . . . ," *Sydney Morning Herald*, July 5–6, 2008.
68. "More than 60 Percent of the US in Drought," Associated Press, July 29, 2006.
69. Tracy Wilkinson, "Spain Withers under Heat Wave and Drought,"

Los Angeles Times, August 8, 2005, http://articles.latimes.com/2005/aug/08/world/fg-drought8.

70. Brian Handwerk, "Scorching Summer Forecast Sees Ten-Degree Rise by 2080," *National Geographic News*, May 10, 2007.

3. The Politics of Environmental Degradation

1. Geoffrey Lean, *Action on Ozone* (Nairobi, Kenya: Information and Public Affairs UNEP, 1989).
2. Greenpeace media release (June 20, 1990), Balmain, New South Wales, Australia, www.greenpeace.org/australia.
3. Jonathon Kwitny, "The Great Transportation Conspiracy," *Harpers*, February 1981, 14–21.
4. Jeremy Leggett, ed., *Global Warming: The Greenpeace Report* (New York: Oxford University Press, 1990).
5. Peter Newman and Jeff Kenworthy, "Greening Urban Transportation," in *State of the World 2007: Our Urban Future*, ed. M. O'Meara (Washington, DC: Worldwatch Institue, 2007).
6. Michael Renner, "Vehicle Production Rises, but Few Cars Are 'Green'" (May 21, 2008), Worldwatch Institute, www.worldwatch.org/node/5461.
7. Marie Woolf and Colin Brown, "Beckett Exposes G8 Rift on Global Warming. Global Warming: The US Contribution in Figures," *Independent* (UK), June 13, 2005, www.commondreams.org/headlines05/0613-02.htm.
8. "Electric Power Monthly" (June 16, 2008), Energy Information Administration, www.eia.doe.gov/cneaf/electricity/epm/table2_3_b.html.
9. Lester R. Brown et al., *State of the World, 1990: A Worldwatch Institute Report on Progress towards a Sustainable Society* (New York: Norton, 1990), 120.
10. "The World Factbook: China" (last updated May 31, 2007), Central Intelligence Agency, www.cia.gov/library/publications/the-world-factbook/geos/ch.html.
11. Catherine Brahic, "China's Emissions May Surpass the US in 2007," *New Scientist*, April 25, 2007.

12. Robert Collier, "The Good Life Means More Greenhouse Gas," *San Francisco Chronicle*, July 6, 2005, www.sfgate.com.
13. "World Energy Demand and Economic Outlook," in International Energy Outlook 2008 (June 2008), Energy Information Administration, www.eia.doe.gov/oiaf/ieo/world.html.
14. Gary Stoller, "Concern Grows over Pollution from Jets," *USA Today*, December 19, 2006.
15. Peter Fairley, "China's Coal Future," *Technology Review*, January 1, 2007, www.technologyreview.com/Energy/18069.
16. Andrew Revkin, "Poorest Nations Will Bear Brunt as World Warms," *New York Times*, April 1, 2007.
17. "China's Signals on Warming," *New York Times*, April 16, 2007.
18. Leggett, *Global Warming*.
19. Walter H. Corson, ed., *The Global Ecology Handbook: What You Can Do about the Environmental Crisis* (Boston: Beacon Press, 1990).
20. Bill Vlasic, "US Pickups Retain a Major Role as Cash Cows," *International Herald Tribune*, January 15, 2008.
21. Bill Vlasic and Nick Bunkley, "Hazardous Conditions," *New York Times*, October 2, 2008.
22. Helen Caldicott, *Missile Envy: The Arms Race and Nuclear War* (New York: Morrow, 1984).
23. Richard Heinberg, "The View from Oil's Peak," *MuseLetter* no. 184 (August 2007), www.richardheinberg.com/museletter/184.
24. Ibid.
25. Robert L. Hirsch, "Peaking of World Oil Production: Impacts, Mitigation, and Risk Management" (February 2005), US Department of Energy, www.netl.doe.gov/publications/others/pdf/Oil_Peaking_NETL.pdf.24A.
26. Heinberg, "View from Oil's Peak."
27. Ibid.
28. Fairley, "China's Coal Future."
29. Ibid.
30. Arjun Makhijani, *Carbon-Free and Nuclear-Free: A Roadmap for U.S. Energy Policy* (Takoma Park, MD: IEER Press, 2007), www.ieer.org.

31. Ibid.
32. Greg Roberts, "The Bad Oil on Ethanol," *Weekend Australian*, May 31–June 1, 2008.
33. Mark Grunwald, "The Clean Energy Scam," *Time*, March 27, 2008.
34. Ibid.
35. Ibid.
36. Ibid.
37. Matthew Wald, "GM Buys Stake in Maker of Corn-Free Ethanol," *International Herald Tribune*, January 15, 2008.
38. Ricardo Radulovich, "Take Biofuels off the Land and Grow Them at Sea" (June 6, 2008), Environmental News Network, www.enn.com/ecosystems/article/37327.
39. Makhijani, *Carbon-Free and Nuclear-Free*.
40. Grunwald, "Clean Energy Scam."
41. "West North Central Household Electricity Report" (March 21, 2006), Energy Information Administration, www.eia.doe.gov.
42. Corson, *Global Ecology Handbook*.
43. Sarah Elks, "Lawn Mower and Outboards New Targets in Emissions War," *Australian*, August 11, 2008.
44. Kevin O'Connor, "Airing His Laundry: Conservationist Pins Hopes on Clotheslines," *Times Argus*, June 22, 2008, www.timesargus.com.
45. Andrew E. Kramer, "Deals with Iraq Are Set to Bring Oil Giants Back," *New York Times*, June 19, 2008.
46. Andrew E. Kramer and Campbell Robertson, "Iraq Cancels Six No-Bid Oil Contracts," *New York Times*, September 11, 2008.
47. Nick Turse, "Pentagon Hands Iraq Oil Deal to Shell" (November 11, 2008), Informationliberation, www.informationliberation.com.
48. Tina Susman, "Poll: Civilian Death Toll in Iraq May Top 1 Million," *Los Angeles Times*, September 14, 2007; Helen Caldicott, *The New Nuclear Danger: George W. Bush's Military-Industrial Complex* (New York: New Press, 2004).
49. Union of Concerned Scientists, *Smoke, Mirrors & Hot Air: How ExxonMobil Uses Big Tobacco's Tactics to Manufacture Uncertainty on*

Climate Science (Cambridge, MA: Union of Concerned Scientists, 2007), www.ucsusa.org/assets/documents/global_warming/exxon_report.pdf.

50. Tom Bergin and Michael Erman, "Rising Oil Prices Power Enormous Exxon Profits," *Truthout*, August 1, 2008.
51. Paul Krugman, "Enemy of the Planet," *New York Times*, April 17, 2006.
52. Ibid.
53. Ibid.
54. Union of Concerned Scientists, *Smoke, Mirrors & Hot Air*.
55. Bergin and Erman, "Rising Oil Prices."
56. "Big Oil, Bigger Giveaways" (July 2008), Friends of the Earth, www.foe.org/pdf/FoE_Oil_Giveaway_Analysis_2008.pdf.
57. Ibid.
58. Helen Caldicott, *Nuclear Power Is Not the Answer* (New York: New Press, 2006).
59. Ibid.
60. *INFACT Brings GE to Light: General Electric, Shaping Nuclear Weapons Policies for Profits* (Boston: INFACT, 1988).
61. Jeff Gerth, "Contractors' Role at Energy Dept. Called Pervasive," *New York Times*, November 6, 1989.
62. Ibid.
63. Caldicott, *Nuclear Power Is Not the Answer*.
64. John M. Broder and Peter Baker, "Obama's Order Is Likely to Tighten Auto Standards," *New York Times*, January 25, 2009.
65. Erica Werner, "Report: EPA Head Reversed Stand on Greenhouse Gas," Associated Press, May 19, 2008.
66. Juliet Eilperin, "Cheney Aides Altered CDC Testimony, Agency Official Says," *Washington Post*, July 9, 2008.
67. Margaret Kriz, "EPA Is Missing in Action on Major Environmental Issues, Observers Charge," *National Journal*, April 14, 2008.
68. Lester R. Brown, *Plan B 3.0: Mobilizing to Save Civilization* (New York: Norton, 2008).
69. James Hansen, "Threat to the Planet: How Can We Avoid Dangerous

Climate Change?" (June 2007), www.apsu.edu/SOARE/index_files/Newsletter.doc.

70. Ibid.
71. Ibid.

4. Trees: The Lungs of the Earth

1. Edward O. Wilson, "Threats to Biodiversity," *Scientific American*, September 1989, 108–16.
2. "Australia's Threatened Species" (March 2006), *NOVA: Science in the News*, Australian Academy of Science, www.science.org.au/.nova/010/010key.htm.
3. Rainforest Live, "Facts and Figures," www.rainforestlive.org.uk/index.cfm?articleid=214.
4. Michael Kennedy, ed., *Australia's Endangered Species: The Extinction Dilemma* (Brookvale, New South Wales: Simon & Schuster Australia, 1990).
5. James Hansen, "Threat to the Planet: How Can We Avoid Dangerous Climate Change?" (June 2007), www.apsu.edu/SOARE/index_files/Newsletter.doc.
6. Nicholas Stern, "Stern Review: The Economics of Climate Change—Executive Summary" (October 2006), www.hm-treasury.gov.uk/d/Executive_Summary.pdf.
7. "Green Carbon: Urgent Need for a Comprehensive Review, National Greenhouse and Energy Reporting (NGER), Comments on Technical Guidelines and Regulations, Margaret Blakers, February 25, 2008," http://climatechange.gov.au/reporting/regulations/pubs/013greeninstitute.pdf.
8. Walter H. Corson, ed., *The Global Ecology Handbook: What You Can Do about the Environmental Crisis* (Boston: Beacon Press, 1990), 235.
9. Ibid., 124.
10. "Eating Up the Amazon" (April 6, 2006), Greenpeace, www.greenpeace.org/usa/assets/binaries/eating-up-the-amazon-executiv; Sting and Jean-Pierre Dutilleux, *Jungle Stories: The Fight for the Amazon* (London: Barrie & Jenkins, 1989), 9, 31.

11. William K. Stevens, "Research in Virgin Amazon Uncovers Complex Farming," *New York Times*, April 3, 1990.
12. Ibid., 71.
13. Rainforest Live, "Facts and Figures."
14. "How Your Tax Dollars Fill the Rainforests" (advertisement by the Rainforest Action Network, San Francisco), *New York Times*, October 15, 1990.
15. "Facts about the Rainforest" (2001–5), Save the Rainforest, www.savetherainforest.org/savetherainforest_007.htm.
16. Eugene Linden, "The Death of Birth," *Time*, January 2, 1989, 32–35.
17. Sting and Dutilleux, *Jungle Stories*, 20.
18. Bruce Rensberger, "Scientists See Signs of Mass Extinction," *Washington Post*, September 29, 1986. The Club of Earth includes Paul Ehrlich, Edward O. Wilson, Ernst Mayer, and Thomas Eisner.
19. "Eating Up the Amazon."
20. Tom Phillips, "The Amazon Burns Once Again," *Guardian* (UK), October 16, 2007.
21. Ian Sample, "Amazon Jungle Could Be Lost in 40 Years, Say Campaigners," *Guardian* (UK), October 2, 2007.
22. "How Your Tax Dollars Fill the Rainforests."
23. Sting and Dutilleux, *Jungle Stories*, 38, 40, 51.
24. Corson, *Global Ecology Handbook*, 124.
25. Sting and Dutilleux, *Jungle Stories*, 15.
26. Petra Kelly, "The Environmental Crisis" (address, Ninth IPPNW World Congress, Hiroshima, Japan, October 9, 1990).
27. Associated Press, "Rainforest Being Wiped Out to Make Chopsticks," *Sydney Morning Herald*, April 15, 1989.
28. "The List: Rising China, Hidden Costs," *Foreign Policy*, March 2007, 11.
29. Rhett A. Butler, "Destruction of Renewable Resources" (January 9, 2006), www.mongabay.com.
30. Sting and Dutilleux, *Jungle Stories*, 39, 41.
31. "Oil and Gas Projects in Western Amazon Threaten Biodiversity

and Indigenous Peoples" (August 13, 2008), Environmental News Network, www.enn.com/ecosystems/article/37915.

32. "The Predicament of Tropical Rainforests: Why They Must Be Saved," Animal Liberation Front, http://animalliberationfront.com/Practical/Health/rainforests.html.
33. Rhett A. Butler, "Tropical Rainforests: Amazon Deforestation. Deforestation in the Amazon" (December 16, 2008), www.mongabay.com.
34. Sting and Dutilleux, *Jungle Stories*, 39.
35. Physicians for Social Responsibility, *Our Common Future, Healing the Planet: A Resource Guide to Individual Action* (Los Angeles: PSR, 1989).
36. "U.S. Links Illegal Drug Production, Environmental Damage. White House Drug Office Notes Deforestation, Pollution, Other Effects" (April 21, 2003), Office of International Information Programs, US Department of State, www.america.gov/st/washfile-english/2003/April/20030421124405retropc0.3930475.html.
37. Sting and Dutilleux, *Jungle Stories*, 43.
38. Douglas Farah, "Cocaine Lords Confound the Amazon Catastrophe," *Australian*, January 12–13, 1991.
39. L. Armstead, "Illicit Narcotics Cultivation and Processing: The Ignored Environmental Drama," United Nations Office on Drugs and Crime, no. 2 (January 1, 1992), www.iisec.ucb.edu.bo/papers/2001-2005/iisec-dt-2002-12.pdf.
40. "Facts about the Rainforest."
41. Sting and Dutilleux, *Jungle Stories*, 33.
42. Ibid.; Wilson, "Threats to Biodiversity"; Corson, *Global Ecology Handbook*, 117, 118.
43. "Facts about the Rainforest."
44. Corson, *Global Ecology Handbook*, 122.
45. "Animal Bytes: Tropical Rain Forest," San Diego Zoo, http://sandiegozoo.org/animalbytes/e-tropical_rainforest.html.
46. Michael Richardson, "South East Asian Nations Move to Protect Forests," *International Herald Tribune*, January 12, 1989.
47. Greg Roberts, "Probe of PNG Logging Bribes," *Australian*, August 20, 2008.

48. Corson, *Global Ecology Handbook*, 228.
49. *State of the World's Forests 2007* (Rome: Food and Agriculture Organization of the United Nations, 2007).
50. Ibid.
51. "Pollution Takes Its Toll on Europe's Forests" (October 11, 2000), Environment News Service, http://forests.org/shared/reader/welcome.aspx?linkid=93851&keybold=forest%20monitoring%20carbon.
52. "Acid Rain" (March 7, 2008), State of Environment Norway, www.environment.no/Tema/Luftforurensning/Sur-nedbor.
53. "Project Water Norway" (July 8, 2007), Norwegian Institute for Water Research, www.water-norway.org/water/proj.niva-proj.html.
54. Corson, *Global Ecology Handbook*, 225, 228; World Commission on Environment and Development, *Our Common Future* (New York: Oxford University Press, 1987), 180.
55. Corson, *Global Ecology Handbook*.
56. "Forest Health Highlights: Forest Resource Summary" (April 2007), US Department of Agriculture, www.fs.fed.us/r5/spf/publications/foresthealth/fhh-ca2006.pdf.
57. "Acid Rain" (February 28, 2006), Kamil's Pollution Web-Site, www.geocities.com/kamil_pollutionpage/AcidRain.
58. "Conservation News," *Newsletter of the Australian Conservation Foundation*, 22, no. 5 (June 1990).
59. "Greening the Ivory Tower" (July 16, 2007), Brandeis University Environmental Studies, www.brandeis.edu/departments/environmental/greeningclassprojects.
60. "Recycling Fun Facts" (July 20, 2007), http://members.aol.com/ramola5/funfacts.html.
61. David McKenzie, "Japanese Consumers Chip in for Forest Protection" (December 15, 2000), Wilderness Society, www.wilderness.org.au/campaigns/forests/consumer/jap_consum.
62. "Disposable Diaper Statistics," Hamptons Diapers, www.hamptonsdiapers.com/disposable-diaper-statistics.html.
63. Allan R. Gold, "Developers' Money Threatening Northern Forests," *New York Times*, April 10, 2005.

64. E. Schwartz, "A Proportionate Mortality Ratio Analysis of Pulp and Paper Mill Workers in New Hampshire," *British Journal of Industrial Medicine* 45 (1988): 234–38.
65. Keith Schneider, "US Backs Off Dioxin's Deadly Ranking as Ecologists Protest," *International Herald Tribune*, August 16, 1991.
66. Ibid.
67. "Dioxin Homepage" (July 11, 2006), EJnet.org, www.ejnet.org/dioxin.
68. "Questions and Answers about Dioxins" (July 7, 2006), US Environmental Protection Agency, http://cfpub.epa.gov/ncea/cfm/recordisplay.cfm/deid=55264&CFID=1356750&CFTOKEN=89850709&jsessionid=66305e33849885148512TR60306630c230.

5. Toxic Pollution

1. Jennifer Gitlitz and Pat Franklin, *Water, Water Everywhere: The Growth of Non-carbonated Beverages in the United States* (Washington, DC: Container Recycling Institute, 2007), www.container-recycling.org/assets/pdfs/reports/2007-waterwater.pdf.
2. Christina Ammon, "Watershed Moment," *High Country News*, June 25, 2007, www.hcn.org/serlets/hcn.
3. Ute Junker, "What's Wrong with a Drink from the Tap?" *Sun Herald*, January 31, 2008.
4. "Bottled Water: Pure Drink or Pure Hype?" Natural Resources Defense Council, www.nrdc.org/water/drinking/bw/bwinx.asp (accessed September 11, 2007).
5. Abid Aslam, "Bottled Water: Nectar of the Frauds?" (February 4, 2006), OneWorld.net, http://us.oneworld.net/article/view/126829/1.
6. Junker, "What's Wrong with a Drink from the Tap?"
7. "Non-hazardous Waste," US Environmental Protection Agency, www.epa.gov/epawaste/basic-solid.htm.
8. Walter H. Corson, ed., *The Global Ecology Handbook: What You Can Do about the Environmental Crisis* (Boston: Beacon Press, 1990), 272.
9. Ibid.

10. Alan B. Durning, "The Quick-Food Addiction," *Arizona Daily Star*, September 17, 1991.
11. William Rathje, "Talkin' Trash," *Washington Post*, February 7, 1999.
12. Cheri Sanders, "City Life: OSC Environmental Medicine Researchers Measure the Risks of Nuclear Living," *USC Medicine*, Winter 1991.
13. Breathe California of Los Angeles County, www.breathela.org.
14. Marla Cone, "State's Air Is among Nation's Most Toxic; Only New York Has a Higher Risk of Cancer Caused by Some Airborne Chemicals, the EPA Says," *Los Angeles Times*, March 22, 2006.
15. "AQMD Launches Major Study of Toxic Air Pollution in Southland" (February 5, 2004), South Coast Air Quality Management District, www.aqmd.gov/news1/2004/matesiiipr.html.
16. "Measuring Vehicle Contribution to Smog," in *Sprawl Report 2001*, Sierra Club, www.sierraclub.org/sprawl/report01/carsandtrucks .asp.
17. Marla Cone, "Farm Air Pollution Targeted; The State Plans Strict, Costly Rules on the Use of Fumigants in Soil," *Los Angeles Times*, May 18, 2007.
18. *Between the Flood and the Rainbow* 4, no. 1&2 (September 1996), Church of the Brethren Network, www.cob-net.org/docs/clean .htm.
19. Marla Cone, "The State; EPA Pesticide Approvals Pose Threat to Species, Report Says; Agency Officials Respond That They Are Strengthening Their Evaluations of the Effects of the Chemicals on Endangered Wildlife," *Los Angeles Times*, July 27, 2004.
20. United Nations Environment Programme (UNEP), "Hazardous Chemicals," UNEP Environment Brief no. 4.
21. "Global Markets for Specialty Pesticides 2006–2007" (August 2007), Kline and Company, www.klinegroup.com/reports/brochures/ y544b/brochure.pdf.
22. "Vital Facts: Selected *Facts and Story Ideas from Vital Signs 2006–2007*" (July 12, 2006), Worldwatch Institute, www .worldwatch.org/node/4346.
23. United Nations Environment Programme (UNEP), "Industry and the Environment," UNEP Environment Brief no. 7.

24. David E. Duncan, "The Pollution Within," *National Geographic*, October 2006.
25. UNEP, "Hazardous Chemicals."
26. Marla Cone, "Common Chemicals are Linked to Breast Cancer," *Los Angeles Times*, May 14, 2007.
27. "Chemical Industry," in *Columbia Encyclopedia*, www.answers.com/topic/chemical-industry.
28. "Vital Facts: Selected Facts and Story Ideas from *Vital Signs 2006–2007*" (July 12, 2006), Worldwatch Institute, www.worldwatch.org/node/4346.
29. "Chemicals Industry of the Future" (May 16, 2007), US Department of Energy, Energy Efficiency and Renewable Energy, www1.eere.energy.gov/industry/chemicals/profile/html.
30. UNEP, "Industry and the Environment."
31. Jim Bell, *Achieving Eco-nomic Security on Spaceship Earth* (San Diego, CA: ELSI Publication, 1995); Tara Bradley-Steck, "Expert: Americans Cause Much Damage to Planet," *Philadelphia Inquirer*, April 6, 1990.
32. "Non-hazardous Waste," US Environmental Protection Agency, www.epa.gov/epawaste/basic-solid.htm; Corson, *Global Ecology Handbook*, 247.
33. Richard Steiner, "While We're Off Fighting Terror, the Planet's Crumbling," *Seattle Post-Intelligencer*, May 30, 2004.
34. "Understanding Radiation: Radioactive Waste" (September 10, 2004), National Safety Council, www.nsc.org/resources/issues/rad/waste.aspx.
35. Rachel L. Gibson, *Toxic Baby Bottles: Scientific Study Finds Leaching Chemicals in Clear Plastic Baby Bottles* (Los Angeles: Environment California Research & Policy Center, 2007), www.environmentcalifornia.org; Marla Cone, "Chemical in Plastics Is Tied to Prostate Cancer," *Los Angeles Times*, June 1, 2006.
36. Marian Burros, "Is There an Extra Ingredient in Nonstick Pans?" *New York Times*, July 27, 2005.
37. Scott Streater, "Nonstick Chemicals May Pose a Threat," *Fort Worth Star-Telegram*, December 5, 2006.
38. "Specialty Stores in Demand for Non-toxic Items like Baby

Organic Clothing," HealthyNewAge.com, www.healthynewage.com/organic-clothing.html.

39. Corson, *Global Ecology Handbook*, 251.
40. John Langone, "A Stinking Mess," *Time*, January 2, 1989, 44–47.
41. Recycling facts (2007), A Recycling Revolution, www.recycling-revolution.com/recycling-facts.html.
42. Layne Nakagawa, "Toxic Trade: The Real Cost of Electronics Waste Exports from the United States" (June 2006), World Resources Institute, http://earthtrends.wri.org/text/population-health/feature-66.html.
43. Corson, *Global Ecology Handbook*, 248.
44. "Cleaning Up Hazardous Waste" (August 2007), NJPIRG, www.njpirg.org/issues/new-jersey-toxics-free-future/cleaning-up-hazardous-waste.
45. William Glaberson, "Love Canal: Suit Centers on Records from 1940s," *New York Times*, October 22, 1990.
46. Ibid.
47. Roxanne Smith, "EPA Adds Seven Sites and Proposes 12 Sites to the Superfund List" (September 19, 2007), US Environmental Protection Agency news release, http://yosemite.epa.gov/opa/admpress.nsf/0/c1c8235c8f2d3ae08525735b0058483b?OpenDocument.
48. "National Priorities List," US Environmental Protection Agency, www.epa.gov/superfund/sites/npl.
49. "Military Toxic Waste Dumping: Killing after the Wars Are Over" (March 8, 2000), Safety Forum, www.safetyforum.com/mtw.
50. Rob Perks, "Pentagon Renews Attack on Public Health and Environment" (March 3, 2005), Natural Resources Defense Council, www.commondreams.org/news2005/0303-05.htm.
51. "Review of Ten Toxic Air Emissions Finds 'Startlingly' Bad Data Reaching Public: Key Flaw: EPA's Failure to Act and New Steps to Undermine Accuracy of Reporting" (June 22, 2004), Environmental Integrity Project, www.commondreams.org/news2004/0622-14.htm.
52. Melissa Cheung, "Overall Pollution Declining" (July 1, 2003), CBS News Online, www.cbsnews.com/stories/2003/07/01/tech/main561237.shtml.

53. "Toxic Pollution and Health: An Analysis of Toxic Chemicals Released in Communities across the United States" (March 22, 2007), U.S. PIRG, www.uspirg.org/home/reports/report-archives/healthy-communities/healthy-communities/toxic-pollution-and-health-an-analysis-of-toxic-chemicals-released-in-communities-across-the-united-states2.
54. Lenny Seigel, Gary Cohen, and Ben Goldman, *U.S. Military's Toxic Legacy: A Special Report on America's Worst Environmental Enemy* (Darby, PA: Diane Publishing, 1991).
55. Mary Tiemann, "Leaking Underground Storage Tanks: Prevention and Cleanup" (January 3, 2007 release), CRS Report for Congress, http://ncseonline.org/NLE/CRs/abstract.cfm?NLEid=1457.
56. Seigel, Cohen, and Goldman, *U.S. Military's Toxic Legacy*.
57. Robert D. Bullard, *Environment and Morality: Confronting Environmental Racism in the United States*, Identities, Conflicts and Cohesion Programme Paper no. 8 (Geneva: United Nations Research Institute for Social Development, 2004).
58. "Cleaning Up Hazardous Waste."
59. Craig E. Colten, "The Rusting of the Chemical Corridor," *Technology and Culture* 47, no. 1 (January 2006), www.historyoftechnology.org/eTC/v47no1/colten.html.
60. Seigel, Cohen, and Goldman, *U.S. Military's Toxic Legacy*.
61. "PVC: The Poison Plastic," Campaign for Safe, Healthy Consumer Products, www.besafenet.com/pvc.
62. Marin Mittelstaedt, "Plastics Ingredient Linked to Smaller Penises," *Globe and Mail*, October 7, 2008.
63. Ibid.
64. "The Impacts of Scrap Electronics & the Benefits of Recycling," Green Star, www.greenstarinc.org/electronicsreasons.php.
65. Jane Houlihan and Richard Wiles, "Lead Pollution at Outdoor Firing Ranges," Environmental Working Group, www.ewg.org/files/leadpoll.pdf.
66. Alexa Olesen, "Mattel Apologises to China over Recalls," Associated Press, September 21, 2007.
67. Marlise Simons, "Nowhere to Hide?" *St. Louis Post-Dispatch*, April 11, 1990.

68. Ibid.
69. Veronika Olaksyn, "WHO: Environment Woes Killing Millions," Associated Press, June 13, 2007, www.boston.com/news/world/europe/articles/2007/06/13/who_environment_woes_killing_millions.
70. Simons, "Nowhere to Hide?"
71. Iva Skochova, "Fight for Life," *Prague Post*, June 28, 2007, www.praguepost.com/articles/2006/06/28/fight-for-life.php.
72. Jochen Klenk, Kilian Rapp, Gisela Büchele, Ulrich Keil, and Stephan K. Weiland, "Increasing Life Expectancy in Germany: Qualitative Contributions from Changes in Age- and Disease-Specific Mortality," *European Journal of Public Health*, February 26, 2007, http://eurpub.oxfordjournals.org/cgi/content/abstract/ckm024v1.
73. Timothy Egan, "A Lonely Law Enforcer Pursues New Violator," *New York Times*, May 19, 1990.
74. Corson, *Global Ecology Handbook*, 253.
75. Ibid.
76. "Pesticide Use in California," PAN Pesticides Database, www.pesticideinfo.org/Search_Use.jsp.
77. Glen Anderson, "Pesticides and Human Health" (July 1999), Environmental Health Series no. 1, National Conference of State Legislatures, www.ncsl.org/programs/environ/envhealth/serpest.htm.
78. Corson, *Global Ecology Handbook*, 252.
79. Keith Addison, e-mail to Biofuel mailing list, January 31, 2001, http://wwia.org/sgroup/biofuel/2456/1.
80. "Race to Go Organic," *Northern Star* (Lismore, Australia), December 19, 1990.
81. Corson, *Global Ecology Handbook*, 253.
82. Ibid., 253–54.
83. "Bhopal Disaster and Aftermath a Huge Violation of Human Rights: Amnesty" (November 29, 2004), Agence France-Presse, www.commondreams.org/headlines04/1129-01.htm.
84. Jim Motavalli, "Arsenic and Old Waste: The Environmental Legacy of Hurricane Katrina," *Environmental Magazine* 27, no. 2 (March–April 2006).

85. Corson, *Global Ecology Handbook*, 250.
86. Basel Convention on the Control of Transboundary Movements of Hazardous Wastes and Their Disposal, www.basel.int.
87. Peter Montague, "Philadelphia Dumps on the Poor" *Rachel's Environment and Health Weekly* no. 595 (April 23, 1998), www.hartford-hwp.com/archives/43a/256.html.
88. Timothy Egan, "Goo Galore," *New York Times*, April 28, 1991.
89. Keith Schneider, "Judge Rejects $100 Million Fine for Exxon in Oil Spill as Too Low," *New York Times*, April 25, 1991.
90. Dianne Solis and Mark Curriden, "Crude Reminders: 19 Years after 'the Day after the Water Died,' Pain of Valdez Spill Still Stings in Alaska," *Dallas Morning News*, March 14, 1999, www.jomiller.com/exxonvaldez/dallas.html.
91. "Exxon Back in Court over 1989 Valdez Spill Fine" (January 27, 2006), MSNBC News Service, www.msnbc.msn.com/id/11059801.
92. Sheila McNulty, "Exxon Valdez Fine Cut by US Supreme Court," *Financial Times*, June 25, 2008, www.ft.com/cms/s/0/04a00cc4-42e6-11dd-81d0-0000779fd2ac.html?nclick_check=1.
93. Thomas C. Hayes, "Earnings Soar 75% at Exxon," *New York Times*, April 25, 1991.
94. Clifford Krauss, "Exxon and Shell Report Record Profits for 2006," *New York Times*, February 2, 2007, www.nytimes.com/2007/02/02/business/02oil.html.
95. David Beers and Catherine Capellaro, "Greenwash," *Mother Jones*, March–April 1991, 38, 39, 40, 41.
96. Fred Montague and Holly Hilton, "Global Energy Consumption: One Wildlife Biologist's Perspective," *Policy Perspectives* 3, no. 8 (August 29, 2007), www.imakenews.com/cppa/e_article000894183.cfm.
97. Paul Brown, "Critical Look in the Mirror for OPEC," *Sydney Morning Herald*, January 29, 1991.
98. "State of the World 2006: China and India Hold World in Balance" (January 11, 2006), Worldwatch Institute, www.worldwatch.org/node/3893.
99. "Making Better Energy Choices," Worldwatch Institute, www.worldwatch.org/node/808.

100. "Fast Facts on Energy Use," Energy Star, www.energystar.gov/ia/business/challenge/learn_more/FastFacts.pdf.
101. "Meeting the Urban Challenge," *Population Reports* 30, no. 4 (Fall 2002), www.infoforhealth.org/pr/m16edsum.shtml.
102. Eric Karlstrom, "Nobody's Hanging Yellow Ribbons for the Persian Gulf," *Sierra Runoff*, April 1991.
103. Barton Gellman, "U.S. Bombs Missed 70% of Time," *Washington Post*, March 16, 1991.
104. *Report to the Secretary-General on Humanitarian Needs in Kuwait and Iraq in the Immediate Post-Crisis Environment*, UN Secretary Council Document S/22366 (March 1991).
105. Alberto Ascherio et al., "Effect of the Gulf War on Infant and Child Mortality in Iraq," *New England Journal of Medicine* 327 (1992): 931–36.
106. "Kuwait, Drilling New Wells, Says Iraq Damage Nears End," *New York Times*, September 15, 1991.
107. John Miller, *Scientists Preview Environmental Effects of a Gulf War* (Brooklyn, NY: International Clearing House on the Military and the Environment, 1991); Michael Parrish, "The Spoils of War," *Los Angeles Times Magazine*, June 23, 1991; Matthew L. Wald, "No Global Threat Seen from Oil Fires," *New York Times*, June 25, 1991.
108. Wald, "No Global Threat."
109. Ibid.
110. "Deadly Black Tide May Be Unstoppable," *Sydney Morning Herald*, January 30, 1991.
111. Deborah Smith, "Oil Threatens Dugongs, Survivors of Another War," *Sydney Morning Herald*, January 30, 1991.
112. Ibid.
113. Sean Ryan, "Oil Spill Poses Threat to Turtles," *Australian*, February 18, 1991.
114. Parrish, "Spoils of War."
115. John Horgan, "Science and the Citizen: Up in Flames," *Scientific American*, May 1991, 24.
116. Ibid., 17.
117. Karlstrom, "Nobody's Hanging Yellow Ribbons."

118. Ibid.
119. Tina Susman, "Poll: Civilian Death Toll in Iraq May Top 1 Million," *Los Angeles Times*, September 14, 2007.
120. Helen Caldicott, *The New Nuclear Danger: George W. Bush's Military-Industrial Complex* (New York: New Press, 2004).
121. Andrew Stephen, "Iraq: The Hidden Cost of War," *New Statesman*, March 12, 2007.
122. Jeff Donn, "Pharmaceuticals Found in US Drinking Water," Associated Press, March 10, 2008.
123. "Wastewater Treatment Plants," Sydney Water, www.sydneywater.com.au/OurSystemsAndOperations/WastewaterTreatmentPlants.
124. "Virus Threatens Mediterranean Dolphins—Spanish Paper" (August 30, 2007), Planet Ark, www.planetark.org/dailynewsstory.cfm?newsid=44014&newsdate=30-Aug-2007.
125. Lucia De Stefano, *Freshwater and Tourism in the Mediterranean* (Rome: WWF Mediterranean Programme, 2004), http://assets.panda.org/downloads/medpotourismreportfinal_ofnc.pdf.
126. Stanley Meisler, "Pollution and Deep Blue Sea," Los Angeles Times, October 20, 1990.
127. Mary-Anne Toy, "Green Games Race against Grime," Sydney Morning Herald, July 8, 2008.
128. "Nuclear Power, Radioactivity," Irish Campaign for Nuclear Disarmament, http://indigo.ie/~goodwill/icnd/power.html.
129. William M. Arkin and Joshua M. Handler, "Nuclear Disasters at Sea, Then and Now," *Bulletin of the Atomic Scientists*, July–August 1989.
130. Caldicott, *New Nuclear Danger*; "Nuclear Weapons around the World," Friends of the Earth, www.motherearth.org/nuke/begin5.php.
131. Corson, *Global Ecology Handbook*, 200.
132. "Tokaimura Criticality Accident, Japan," in *Encyclopedia of Earth*, www.eoearth.org/article/Tokaimura_criticality_accident,_Japan.
133. Helen Caldicott, *Nuclear Power Is Not the Answer* (New York: New Press, 2006); Caldicott, *New Nuclear Danger*; Susan Wyndham, "Death in the Air," *Australian Magazine*, September 29–30, 1990; Keith Schneider, "Opening the Record on Nuclear Risks," *New York Times*, December 3, 1989; Kenneth B. Noble, "The US for

Decades Let Uranium Leak at Weapon Plant," *New York Times*, October 15, 1988.

134. Wyndham, "Death in the Air."
135. Ibid.
136. Matthew L. Wald, "Chemicals in a Plant Pose Blast Risk," *New York Times*, October 17, 1989.
137. Helen Caldicott, *Nuclear Madness: What You Can Do* (New York: Bantam, 1980), 42.
138. Martha Odom, "Tanks That Leak, Tanks That Explode . . . Tanks Alot DOE," *Portland Free Press*, May 1989.
139. Matthew L. Wald, "Wider Peril Seen in Nuclear Waste from Bomb Making," *New York Times*, March 28, 1991.
140. Wyndham, "Death in the Air."
141. Larry Lang, "Missing Hanford Documents Probed by Energy Department," *Seattle Post-Intelligencer*, September 20, 1991.
142. "Nuclear Waste Spill Linked to Water Line" (August 2, 2007), United Press International, www.upi.com/Top_News/2007/08/02/Nuclear_waste_spill_linked_to_water_line/UPI-11771186031485.
143. Wyndham, "Death in the Air."
144. Keith Schneider, "Seeking Victims of Radiation near Weapon Plant," *New York Times*, October 17, 1988.
145. James Long, "A Tour of Hanford Reveals the Dangers of the Birthplace of the Bomb," *Oregonian*, June 28, 2008.
146. Noble, "US for Decades Let Uranium Leak."
147. Schneider, "Opening the Record."
148. Keith Schneider, "Brain Cancer Cases in Los Alamos to Be Studied for Radiation Link," *New York Times*, July 23, 1991.
149. Matthew L. Wald, "In Nuclear Cleanup, Costs Grow and Promises Fade," *New York Times*, October 29, 1989.
150. Brad Knickerbocker, "Huge Cleanup Awaits Arms Plants," *Christian Science Monitor*, March 15, 1991.
151. Matthew L. Wald, "The Adventures of Toxic Avengers Have Barely Begun," *New York Times*, September 15, 1991.
152. "Nuclear Power Plants Information: Number of Reactors in Operation Worldwide," International Atomic Energy Agency, www

.iaea.org/cgi-bin/db.page.pl/pris.oprconst.htm (accessed January 15, 2009).

153. Corson, *Global Ecology Handbook.*

154. John Vidal, "Hell on Earth," *Guardian* (UK), April 26, 2006.

155. "Chernobyl Children's Charities Spread Dangerous Myths" (March 2006), Chernobyl Legacy, www.chernobyllegacy.com/index.php?cat.

156. "Chernobyl—The Facts: What You Need to Know Almost 20 Years after the Disaster," Chernobyl Children's Project International, www.chernobyl-international.org.

157. "Chernobyl—The Facts."

158. Linda Gunter and Paul Gunter, "Chernobyl Can Happen Here" (May 25, 2005), Minuteman Media, www.commondreams.org/views05/0525-26.htm.

159. "Chernobyl: The Facts."

160. "Chernobyl Update" (radio broadcast, April 30, 1991), KPFA (Los Angeles).

161. "Risk Assessment of Chernobyl Accident Consequences: Lessons Learned for the Future" (2005), NATO/CCMS Pilot Study, www.nato.int/science/pilot-studies/raca/041204a.pdf.

162. Jennifer del Rosario-Malonzo, "US Military-Industrial Complex: Profiting from War" (2002), IBON Features 2002-22, http://nadir.org/nadir/initiativ/agp/free/9-11/military_complex.htm.

163. Jan Willem Storm van Leeuwen and Philip Smith, "Nuclear Power, the Energy Balance" (January 2008), http://beheer.opvit.rug.nl/deenen/Nuclear_sustainability_rev3.doc.

6. Species Extinction

1. Julia Whitty, "By the End of the Century Half of All Species Will Be Gone. Who Will Survive?" *Mother Jones*, May–June 2007.

2. Edward O. Wilson, "Threats to Biodiversity," *Scientific American*, September 1989, 108–16.

3. Dan Olson, "Species Extinction Rate Speeding Up" (radio broadcast, February 1, 2005), Minnesota Public Radio.

4. Whitty, "By the End of the Century."

5. "The Review of the 2008 Red List of Threatened Species," International Union for Conservation of Nature, www.iucn.org/about/work/programmes/species/red_list/review.
6. "Review of the 2008 Red List."
7. Ibid.
8. Wilson, "Threats to Biodiversity."
9. Ibid.
10. World Commission on Environment and Development, *Our Common Future* (New York: Oxford University Press, 1987), 4.
11. Andrew Downey, "Amazon Harvest," *Nature Conservancy* 57, no. 3 (Autumn 2007): 38–41.
12. Joy B. Zedler, "Compensating for Wetland Losses in the United States," *Ibis*, 146, no. s1 (September 2004): 92–100.
13. Emilie C. Snell-Rood and Daniel A. Cristol, "Avian Communities of Created and Natural Wetlands: Bottomland Forests in Virginia," *Condor*, 105 (May 2003): 303–315.
14. *Climate Change 2007: Impacts, Adaptation and Vulnerability: Contribution of Working Group II to the Fourth Assessment Report of the Intergovernmental Panel on Climate Change* (Cambridge: Cambridge University Press, 2007).
15. John Garnaut, "Goldminer Pulls Plug on Pursuit of Summer," *Sydney Morning Herald*, June 7–8, 2008.
16. "Review of the 2008 Red List."
17. Greg Roberts, "Frogs Back from the Dead," *Australian*, May 26, 2008.
18. Richard Cole, "Amphibians in Global Decline Say Scientists," *Northern Star* (Lismore, Australia), June 20, 1990; Stan Ingram, "Of Fire, Water, Earth and Air—The Mystery of the Disappearing Frog," *Wildlife*, Spring 1990.
19. Ibid.
20. "Fifty Key Facts about Seas and Oceans," Ocean Foundation, http://archive.oceanfdn.org/index.php?tg=articles&idx=Articles&topics=32#art155.
21. J. Hansen, R. Ruedy, M. Sato, and K. Lo, *Global Temperature Trends: 2005 Summation* (New York: NASA Goddard Institute for Space Studies, 2005), http://data.giss.nasa.gov/gistemp/2005.

22. "Warming Climate Linked to Reef Destruction" (December 6, 2004), Environment News Service, www.ens-newswire.com/ens/dec2004/2004-12-06-01.asp.
23. "Fifty Key Facts about Seas and Oceans."
24. Ibid.
25. Michael Kennedy, "Endangered," *Habitat*, August 29, 1990.
26. John Nicholson, *The State of the Planet* (St. Leonards, New South Wales, Australia: Allen & Unwin, 2000), 13.
27. Roger Beckman and Steven Morten, "Where Have All the Desert Mammals Gone?" *Wildlife*, Spring 1990.
28. "Review of the 2008 Red List."
29. "African Elephants," World Wildlife Fund, www.panda.org/about_wwf/what_we_do/species/about_species/species_factsheets/elephants/african_elephants.
30. Walter H. Corson, ed., *The Global Ecology Handbook: What You Can Do about the Environmental Crisis* (Boston: Beacon Press, 1990), 103.
31. Mark Henderson, "23,000 Elephants Killed Each Year for Ivory," *Times Online*, February 27, 2007, www.timesonline.co.uk/tol/news/uk/science/article1443967.ece.
32. "African Elephants—Threats: Still Poached for Meat and Ivory" (February 12, 2007), World Wildlife Fund, www.panda.org/about_wwf/what_we_do/species/about_species/species_factsheets/elephants/african_elephants/afelephants_threats/index.cfm.
33. Damien Cave, "Everglades Deal Now Only Land, Not Assets," *New York Times*, November 12, 2008.
34. Richard Macey, "Terminal Diagnosis for Ocean Creatures," *Sydney Morning Herald*, November 12, 2008.
35. "Blue Whale" (February 13, 2007), World Wildlife Fund, www.panda.org/about_wwf/what_we_do/species/about_species/species_factsheets/cetaceans/blue_whale.
36. "Whaling," Greenpeace, http://oceans.greenpeace.org/en/our-oceans/whaling.
37. "Mammals," in *Living Library* (Wildwatch.com), www.wildwatch.com/living_library/mammals-2/white-rhinoceros.

38. Jeffrey Gettleman, "Congo Violence Reaches Endangered Mountain Gorillas," *New York Times*, November 18, 2008.

39. "Humans Driving Primates to the Brink," Associated Press, August 6, 2008.

40. Nick Galvin, "Boy's Own Passion for Plight of the Big Cats," *Sydney Morning Herald*, August 20, 2008.

41. David Charley, "Famous Finches Feel the Pressure," *Northern Star* (Lismore, Australia), June 30, 1990.

42. Jane Brody, "Studies Point to Food Web Danger," *Sydney Morning Herald*, February 15, 1989; *A Strategy for Antarctic Conservation* (Proceedings of the Eighteenth Session of the IUCN General Assembly, Perth, Australia, November 28–December 5, 1990).

43. "Human Impacts on Antarctica and Threats to the Environment—Fishing," Cool Antarctica, www.coolantarctica.com/Antarctica%20fact%20file/science/threats_fishing_hunting.htm.

44. Maricela Yip, "Ozone in the Atmosphere" (December 1, 2000), www.sbg.ac.at/ipk/avstudio/pierofun/atmo/ozone.htm.

45. "Antarctic Ozone," British Antarctic Survey, www.antarctica.ac.uk/met/jds/ozone.

46. "Wilkins Ice Shelf at Risk of Breaking from Antarctic" (July 12, 2008), ABC News, www.abc.net.au/news/stories/2008/07/12/2301885.htm.

47. "Massive Ice Shelf 'May Collapse without Warning,'" *New Zealand Herald*, December 1, 2006.

48. "Legislative Instrument Details: Madrid Protocol" (January 14, 1998), Eionet, http://rod.eionet.europa.eu/instruments/576.

49. "Customs Australia Finds Rare Lizard, Turtles in Mail" (March 26, 2004), Environment News Service, www.ens-newswire.com/ens/mar2004/2004-03-26-01.asp.

50. "The Bees' Needs," Natural Resources Defense Council, www.nrdc.org/wildlife/animals/bees.asp?gclid.

51. Al Meyerhoff, "Buzzzzzzzz Kill: The Loss of Billions of Bees Raises Questions about Our Pesticide Controls," *Los Angeles Times*, July 30, 2008.

52. Simon Webster, "Bee Story Has Nasty Sting in the Tail," *Sun-Herald*, July 6, 2008.

53. Carolyn Lockhead, "Farm Bill Little Help to Plight of Bees; Subsidies Expected to Result in More Deaths of Pollinators," *San Francisco Chronicle*, April 19, 2008.
54. Elizabeth R. Dumont, "Feeding Mechanisms in Bats: Variation within the Constraints of Flight," *Integrative and Comparative Biology* 47 (2007): 137–146, icb.oxfordjournals.org/cgi/content/full/47/1/137.
55. "GIWA Regional Assessment 17—Baltic Sea," United Nations Environment Programme, www.unep.org/dewa/giwa/areas/reports/r17/assessment_giwa_r17.pdf.
56. Peter Pringle, "The Green Bottles Don't Accidentally Fall on a US Heap," *Sydney Morning Herald*, July 28, 1989.
57. Patty Conklin Steele, personal communication, October 24, 2007.
58. David Adams, "Suffocating Dead Zones Spread across World's Oceans," *Guardian* (UK), August 15, 2008.
59. "WWF International Smart Gear Competition Hooks Multinational Experts to Choose Winner" (February 15, 2006), World Wildlife Fund, www.worldwildlife.org/who/media/press/2006/WWFPresitem835.html.
60. "Fishing Nets Major Risk for Small Cetaceans" (November 23, 2005), United Nations Environment Programme, www.unep.org/Documents.Multilingual/Default.asp?DocumentID=457&ArticleID=5053&I=en.
61. Elisabeth Rosenthal, "Appetite for Seafood Puts Fish Stocks in Peril," *International Herald Tribune*, January 16, 2008.
62. Sea World, Gold Coast, Australia, "Drift Net Fishing," Environmental Issues Booklet 2, http://myfun.com.au/images/temppdf/sw/education_brochure_pdfs/fwenvironmentalissues/Environmental%20Issues%20Booklet%202.pdf.
63. Ibid.
64. "Bycatch," Greenpeace, www.greenpeace.org/international/campaigns/oceans/bycatch.
65. Paul Grigson and Judith Whelan, "Japan Bows to Pressure against Drift Net Fishing in the Southern Pacific," *Sydney Morning Herald*, July 18, 1990.
66. "Whales, Dolphins and Porpoises," Australian Government, Depart-

ment of the Environment, Water, Heritage and the Arts, www.environment.gov.au/coasts/species/cetaceans/dolphins.html.

67. Elisabeth Rosenthal, "Stinging Tentacles Offer Hint of Oceans' Decline," *New York Times*, August 3, 2008.
68. Brian Woodley, "Moratorium Fails to Stop Killing of Whales," *Weekend Australian*, June 30–July 1, 1990.
69. "The Key to Ending Whaling: Changing Perceptions in Japan" (February 19, 2007), Greenpeace, http://oceans.greenpeace.org/en/the-expedition/news/the-key-to-ending-whaling.
70. "Whale Conservation—Protecting the Giants of the Sea," WWF-Australia, www.wwf.org.au/ourwork/oceans/whales.
71. "Russia Defies IWC: Allows Chukotka Inuit to Take Two Bowheads," *High North News*, no. 11 (November 1996), www.highnorth.no/library/Hunts/Other/ru-de-iw.htm.
72. "Whales, Dolphins, Sonar and the Courts," *New York Times*, August 10, 2008.
73. Dianne Dumanowski, "Measles-like Virus Reported in Dolphins," *Boston Globe*, October 24, 1990.
74. Ibid.
75. Timothy Egan, "Environmentalists Flipping over Dolphin Use as Navy Guard Dogs," *Sydney Morning Herald*, April 15, 1989.
76. Thomas Watkins, "Navy May Deploy Anti-Terrorism Dolphins," *USA Today*, February 13, 2007, www.usatoday.com/tech/science/2007-02-13-dolphin-defenders_x.htm.
77. Egan, "Environmentalists Flipping."
78. Naval Oceans Systems Center, San Diego, California, personal communication.
79. Maggie Tieger, "Species Conservation under Appendix 1," *Endangered Species Bulletin* 30, no. 2 (September 2005), 23.

7. Overpopulation

1. "Negative Population Growth: Facts and Figures: Total Midyear World Population, 1950–2050," US Bureau of the Census, http://npg.org/facts/world_pop_year.htm.

2. "World Population to Reach 9.1 Billion by 2050, UN Projects" (February 24, 2005), UN News Centre, www.un.org/apps/news/story.asp?NewsID=13451&Cr=population&Cr1.
3. Nils Blythe, "India's Big Population Challenge" (February 24, 2008), BBC News, http://news.bbc.co.uk/2/hi/business/7261458.stm.
4. Thomas H. Tietenberg's personal page at Colby College, topic: population control of India, www.colby.edu/personal/t/thtieten/Famplan.htm.
5. Dhananjay Mahapatra, "India Misses Population Control Targets for 2010, 2016," *Times of India*, March 27, 2008, http://timesofindia.indiatimes.com/India/India_misses_population_control_targets_for_2010_2016/articleshow/3220037.cms.
6. Matt Rosenber, "China Population: The Population Growth of the World's Largest Country" (March 4, 2008), About.com, http://geography.about.com/od/populationgeography/a/chinapopulation.htm.
7. Sanjay Suri, "Women: Half the Population, A Fifth of the News" (February 15, 2006), Inter Press Service, www.commondreams.org/headlines06/0215-10.htm.
8. "About the Goals," Millennium Campaign, http://millenniumcampaign.org/site/pp.asp?c=grKVL2NLE&b=186382.
9. "What Is Child Survival?" US Coalition for Child Survival, www.child-survival.org/childsurvival/whatiscs.cfm.
10. Anup Shah, "Poverty Facts and Figures," Global Issues, www.globalissues.org/TradeRelated/Facts.asp.
11. Ibid.
12. Gilda Sedgh, "Millions of Women at Risk of Unplanned Pregnancy in Developing Nations Are Not Using Contraceptives" (July 2007), Population and Health InfoShare, www.phishare.org/documents/Guttmacher/5028.
13. Robert Steinbrook, "Gains Reported in Test of Male Contraceptive," *Los Angeles Times*, October 20, 1990.
14. "The Bush Global Gag Rule: Endangering Women's Health, Free Speech and Democracy" (July 1, 2003), Center for Reproductive Rights, http://crr.civicactions.net/en/document/the-bush-global-gag-rule-endangering-womens-health-free-speech-and-democracy.

15. Rebecca Buckwalter-Poza, "Reproductive Rights without Borders," *Campus Progress*, January 4, 2007, http://campusprogress.org/features/1351/reproductive-rights-without-borders.
16. L. Silvestre, C. Dubois, M. Renault, Y. Rezvani, E. E. Baulieu, and A. Ulmann, "Voluntary Interruption of Pregnancy with Mifepristone (RU 486) and a Prostaglandin Analogue. A Large-Scale French Experience," *New England Journal of Medicine* 322 (March 8, 1990), 645–48.
17. Fred Kaplan, "Our Hidden WMD Program: Why Bush Is Spending So Much on Nuclear Weapons," *Slate*, April 23, 2004, www.slate.com/id/2099425.
18. Anup Shah, "World Military Spending," Global Issues, www.globalissues.org/Geopolitics/ArmsTrade/Spending.asp.
19. Walter H. Corson, ed., *The Global Ecology Handbook: What You Can Do about the Environmental Crisis* (Boston: Beacon Press, 1990), 52.
20. "Abortion Worldwide," Pro-Life Infonet, www.euthanasia.com/globe.html.
21. Cuban consul general (Sydney, Australia), personal communication, October 1991.
22. "Increasing Diversity Predicted in U.S. Population: Hispanic and Asian Populations Both Poised to Triple" (March 18, 2004), US Department of State, Bureau of International Information Programs, www.america.gov/st/washfile-english/2004/March/20040318124311CMretroP0.4814264.html.
23. Corson, *Global Ecology Handbook*, 52.
24. Ibid., 53.

8. First World Greed and Third World Debt

1. F. E. Trainer, *Abandon Affluence!* (London: Zed Books, 1985), 116.
2. "The Global Rich and Poor Gap Widens" (January 24, 2007), Voice of America, www.voanews.com/english/archive/2007-01/GlobalRichandPoor2006-12-22-voa6.cfm?CFID=151111445&CFTOKEN-61268573.
3. "U.S. Life Expectancy May Drop Due to Obesity," Associated Press, March 16, 2005, www.msnbc.msn.com/id/7209499.

4. "Hunger Facts: International," Bread for the World, www.bread.org/learn/hunger-basics/hunger-facts-international.html.
5. "Half of US Food Goes to Waste" (November 25, 2004), Food ProductionDaily, www.foodproductiondaily.com/Supply-Chain/Half-of-US-food-goes-to-waste.
6. Francisco Sevilla, "Childhood Obesity: The Greatest Risk for America's Children," Insight Business, www.usc.edu/org/Insight Business/SP07articles/obesity.html.
7. Matthew Christensen, "Malnourishment in the United States," www.worldfoodprize.org/assets/YouthInstitute/05proceedings/Jefferson-ScrantonHighchool.pdf.
8. David W. Boles, "Born Poor and Condemned to Lifelong Poverty" (August 9, 2007), Urban Semiotic, http://urbansemiotic.com/2007/08/09/born-poor-and-condemned-to-lifelong-poverty.
9. Walter H. Corson, ed., *The Global Ecology Handbook: What You Can Do about the Environmental Crisis* (Boston: Beacon Press, 1990), 72.
10. "Long-term High Consumption of Red and Processed Meat Linked with Increased Risk for Colon Cancer," *ScienceDaily*, January 13, 2005, www.sciencedaily.com/releases/2005/01/050111164223.htm.
11. Christian Nordqvist, "Heavy Red Meat Consumption Raises Breast Cancer Risk Considerably," *Medical News Today*, November 14, 2006, http://medicalnewstoday.com/articles/56707.php.
12. "Livestock," in *Science Encyclopedia*, http://science.jrank.org/pages/3980/Livestock.html.
13. Peter Singer and Jim Mason, "Eating" (September 7, 2006), BBC Newsnight, http://news.bbc.co.uk/2/hi/programmes/newsnight/5313424.stm.
14. "U.S. Could Feed 800 Million People with Grain That Livestock Eat, Cornell Ecologist Advises Animal Scientists," *ScienceDaily*, August 12, 1997, http://sciencedaily.com/releases/1997/08/970812003512.htm.
15. Corson, *Global Ecology Handbook*, 28, 73, 77.
16. "U.S. Could Feed 800 Million People."
17. Danielle Murray, "Oil and Food: A New Security Challenge," *Asian Times*, June 3, 2005, www.atimes.com/atimes/Global_Economy/GF03Dj01.html.

18. Frosty Wooldridge, "Food and Environment: Part 26—Next Added 100 Million Americans," *American Chronicle*, May 21, 2007, www.americanchronicle.com/articles/viewArticle.asp?articleID=27686.
19. Judy Putnam, Jane Allshouse, and Linda Scott Kantor, "U.S. Per Capita Food Supply Trends: More Calories, Refined Carbohydrates, and Fats," *FoodReview*, Winter 2005, www.ers.usda.gov/publications/FoodReview/DEC2002/frvol25i3a.pdf.
20. Trainer, *Abandon Affluence!*, 157.
21. Ian Sample, "Global Food Crisis Looms as Climate Change and Population Growth Strip Fertile Land," *Guardian* (UK), August 31, 2007, http://guardian.co.uk/environment/2007/aug/31/climatechange.food.
22. David Pimentel and Nadia Kounang, "Ecology of Soil Erosion in Ecosystems," *Ecosystems* 1 (September 1, 1998), 416–426.
23. "The United States and International Development: Partnering for Growth" (August 6, 2007), US Department of State, www.state.gov/r/pa/prs/ps/2007/aug/90348.htm.
24. Claudia Parson, "US Aid Should Tackle Poverty before Security: Report," *Boston Globe*, October 18, 2007.
25. Anup Shah, "World Military Spending," Global Issues, www.globalissues.org/Geopolitics/ArmsTrade/Spending.asp.
26. Trainer, *Abandon Affluence!*, 177.
27. Vexen Crabtree, "Foreign Aid" (August 23, 2008), www.vexen.co.uk/USA/foreign_aid.html.
28. Trainer, *Abandon Affluence!*, 146, 157; Corson, *Global Ecology Handbook*, 57.
29. "2006 Box Office Rebounds: Global Box Office Reaches Historic High" (March 6, 2007), Motion Picture Association of America, www.mpaa.org/press_releases/2006%20market%20stats%20release%20final.pdf.
30. "Economics of Alcohol and Tobacco—U.S. Tobacco Production and Consumption," Table 7.6: Expenditures for Tobacco Products and Disposable Personal Income, 1993–2003, www.libraryindex.com/pages/2129/Economics-Alcohol-Tobacco-U-S-TOBACCO-PRODUCTIONAND-CONSUMPTION.html.

31. "Alcohol Facts You'll Never Hear from Big Booze" (April 2006), Fetal Alcohol Disorders Society, www.acbr.com/fas/alcaware.htm.
32. "Diet Industry Is Big Business: Americans Spend Billions on Weight-Loss Products Not Regulated by the Government" (December 1, 2006), CBS Evening News, www.cbsnews.com/stories/2006/12/01/eveningnews/main2222867.shtml.
33. John P. Whitcher, M. Srinivasan, and Madan P. Upadhyay, "Corneal Blindness: A Global Perspective," *Bulletin of the WHO* 79, no. 3 (2001).
34. Trainer, *Abandon Affluence!*, 177.
35. "iPod—A Grassroots Project" (December 12, 2006), Some Random Dude, www.somerandomdude.net/blog/design/grassroots-project-one-ipod.
36. Anup Shah, "US and Foreign Aid Assistance," Global Issues, www.globalissues.org/TradeRelated/Debt/USAid.asp.
37. Photius Coutsoukis, "Literacy—Total (%) 2007" (January 2007), www.photius.com/rankings/population/literacy_total_2007_0.html.
38. Jeff Green, "U.S. Has Second Worst Newborn Death Rate in Modern World, Report Says" (May 10, 2006), CNN, www.cnn.com/2006/HEALTH/parenting/05/08/mothers.index/index.html.
39. Shah, "World Military Spending."
40. William D. Hartung, "Soldiers versus Contractors: Emerging Budgetary Reality?" (February 10, 2006), World Policy Institute, http://worldpolicy.org/projects/arms/reports/soldiers.html.
41. Alameda County SANE/FREEZE, "Taxes" (1987).
42. Joseph Stiglitz and Linda Bilmes, "The Three Trillion Dollar War: The Cost of the Iraq and Afghanistan Conflicts Have Grown to Staggering Proportions," *Times* (UK), February 22, 2007, www.timesonline.co.uk/tol/comment/columnists/guest_contributors/article3419840.ece; William D. Hartung, "War Is Hell, But What the Hell Does It Cost?" (March 5, 2008), TomDispatch.org, www.commondreams.org/archive/2008/03/05/7477.
43. Helen Caldicott, *Missile Envy: The Arms Race and Nuclear War* (New York: Morrow, 1984).
44. Paulo Nakatani and Rémy Herera, "The South Has Already Repaid

Its External Debt to the North," *Monthly Review* 59, no. 2 (June 2007), www.monthlyreview.org/0607pnrh.html.

45. Susan George, "A Fate Worse than Debt" (BBC documentary), 1990.
46. Steve Hargreaves, "Oil Slips Despite Supply Drop, No OPEC Hike" (December 5, 2007), CNNMoney.com, http://money.cnn.com/2007/12/05/markets/oil_eia.
47. Wayne Arnold, "World Business Briefing: Asia: Economic Aid for Philippines," *New York Times*, June 21, 2000.
48. "Philippines: Poverty and Wealth," in *Encyclopedia of the Nations*, www.nationsencyclopedia.com.
49. George, "Fate Worse than Debt."
50. "Annual Report 2007: Latin America and the Caribbean," World Bank, www.worldbank.org.
51. David R. Francis, "The Mystery of the Missing $2.9 Trillion," *Christian Science Monitor*, October 29, 2007, www.csmonitor.com/2007/1029/p15s01-wmgn.html.
52. Anup Shah, "Third World Debt Undermines Development," Global Issues, www.globalissues.org/TradeRelated/Debt.asp.
53. Bobby Webster, "Developing World Debt" (last modified June 20, 2005), International Debate Education Association, www.idebate.org/debatabase/topic_details.php?topicID=45.
54. James R. Hagerty and Deborah Solomon, "Take a Load off Fanny," *Australian/Wall Street Journal*, July 16, 2008.
55. George, "Fate Worse than Debt."
56. "Can New Loan Really Bring Sustainable Cattle Ranching to the Amazon?" (March 12, 2007), Mongabay.com, http://news.mongabay.com/2007/0312-beef.html.
57. "Investing in Destruction—The World Bank and Biodiversity" (November 1996), GRAIN, www.grain.org/briefings/?id=34.
58. John Vidal, "World Bank Accused of Razing Congo Forests," *Guardian* (UK), October 4, 2007.
59. Carol Sherman, *A Look inside the World Bank* (Sydney, Australia: Envirobook, 1990).
60. Waldon Bello, "Manufacturing a Food Crisis," *Nation*, June 2, 2008.

61. Corson, *Global Ecology Handbook*, 45, 46.
62. Naomi Klein, *The Shock Doctrine: The Rise of Disaster Capitalism* (New York: Metropolitan Books, 2007), 163.
63. Ibid., 457.
64. George, "Fate Worse than Debt."
65. Trainer, *Abandon Affluence!*, 141.
66. "U.S. Food Imports Rarely Inspected," Associated Press, April 16, 2007, www.msnbc.msn.com/id/18132087.
67. Fantu Cheru, "Development, Debt and Dependency," Multi-National Monitor, http://multinationalmonitor.org/hyper/issues/1988/07/mm0788_05.html.
68. "Poverty in Africa," in *Wikipedia*, http://en.wikipedia.org/wiki/Poverty_in_Africa.
69. George, "Fate Worse than Debt."
70. "Forest Destruction for Export," *WRM Bulletin* no. 85 (August 2004), www.wrm.org.uy/bulletin/85/LA.html.
71. "Country Profile: Venezuela" (March 2005), Mongabay.com, www.mongabay.com/reference/country_profiles/2004-2005/Venezuela.html.
72. "Nicaragua: Food Security Update" (December/January 2006/07), MFEWS, http://v4.fews.net/docs/Publications/Nicaragua_200611en.pdf.
73. Matthew Clark, "In Trademarking Its Coffee, Ethiopia Seeks Fairness," *Christian Science Monitor*, November 9, 2007, www.csmonitor.com/2007/1109/p01s06-woaf.html.
74. "India: Reinterpreting Tea Leaves," MercyCorps, www.mercycorps.org/countries/india.
75. Trainer, *Abandon Affluence!*, 143.
76. Ibid., 153.
77. "Mexico: Facts," WorldPress.org, http://worldpress.org/profiles2/Mexico.cfm.
78. Chris Clarke, "Fat and Greed" (June 5, 2006), Creek Running North, http://faultline.org/index.php/site/comments/fat_and_greed.
79. James Petras, "Mexico Is a Virtual Trade Colony of the US," Foreign Exchange, www.zmag.org/ZMag/articles/petrasapr98.htm.

80. Jacob Hill, "Free Trade and Immigration: Cause and Effect" (July 18, 2007), Council on Hemispheric Affairs, www.coha.org/2007/07/free-trade-and-immigration-cause-and-effect.

81. "Globalization: Corporate Earnings Soar as Wage Growth Stalls in Developed World; Wages Rising in China" (April 3, 2006), Finfacts, www.finfacts.com/irelandbusinessnews/publish/printer_1000article_10005404.shtml.

82. Mingan Choct, "Role of Biotechnology in Utilisation of Alternative Feed Ingredients for Monogastric Animals" (March 16, 2007), Engormix, http://engormix.com/role_of_biotechnology_in_e_articles_397_BAL.htm.

83. Rosemary Lugg, Louise Morley, and Fiona Leach, *Country Profiles for Ghana and Tanzania: Economic, Social and Political Contexts for Widening Participation in Higher Education* (University of Sussex, University of Cape Coast, University of Dar es Salaam, 2007), www.sussex.ac.uk/education/documents/country_profiles_24may_2007.pdf.

84. Alan Cowell, "Finance Chiefs Cancel Debt of 18 Nations," *New York Times*, June 12, 2005, www.nytimes.com/2005/06/12/international/12debt.html.

85. Alex Wilks, "Selling Africa Short" (June 20, 2005), Open Democracy, www.opendemocracy.net/globalization-G8/debt_2616.jsp.

86. Corson, *Global Ecology Handbook*, 46–47.

87. George, "Fate Worse than Debt."

88. Ibid.

89. Corson, *Global Ecology Handbook*, 47.

90. Alberto Ramos, "The Secretary-General's Agenda: A Unique Opportunity in Latin America for Reform and Growth," *UN Chronicle*, www.un.org/Pubs/chronicle/2007/webArticles/032907_latinamerica.htm.

91. Mark Carlson and Gretchen Weinbach, "Profits and Balance Sheet Developments at U.S. Commercial Banks in 2006," *Federal Reserve Bulletin* 93 (April 13, 2007), www.federalreserve.gov/pubs/bulletin/2007/pdf/bankprofit07.pdf.

92. "Brazil's Annual Inflation Rate Reaches 4.12%," *Pravda*, November 7, 2007, http://english.pravda.ru/news/business/07-11-2007/100401-inflation_rate-0.

93. Mac Margolis, "How Brazil Reversed the Curse," *Newsweek*, November 12, 2007, www.newsweek.com/id/67850.

94. Albert Einstein, quoted in the *New York Times*, May 25, 1946.

95. M. Boko, I. Niang, A. Nyong, C. Vogel, A. Githeko, M. Medany, B. Osman-Elasha, R. Tabo, and P. Yanda, "2007: Africa," in *Climate Change 2007: Impacts, Adaptation and Vulnerability: Contribution of Working Group II to the Fourth Assessment Report of the Intergovernmental Panel on Climate Change* (Cambridge: Cambridge University Press, 2007), 433–67.

96. "Massive Starvation Looms in Africa," *Northern Star* (Lismore, Australia), December 19, 1990.

97. *Climate Change 2007: Impacts, Adaptation and Vulnerability: Contribution of Working Group II to the Fourth Assessment Report of the Intergovernmental Panel on Climate Change* (Cambridge: Cambridge University Press, 2007).

98. "Executive Summary: Population Dynamics and Global Climate Change" (2004), Population Resource Center, www.prcdc.org/files/2002%20Population%20Dynamics%20and%20Global%20Climate%20Change.

99. "Massive Starvation Looms in Africa."

100. Corson, *Global Ecology Handbook*, 68.

101. Lester R. Brown, "China's Shrinking Grain Harvest: How Its Growing Grain Imports Will Affect World Food Prices" (March 10, 2004), Earth Policy Institute, www.earth-policy.org/Updates/Update36.htm.

102. John Vidal and Tim Radford, "One in Six Countries Facing Food Shortage," *Guardian* (UK), June 30, 2005, www.guardian.co.uk/world/2005/jun/30/science.famine.

103. "World Day to Combat Desertification," International Fund for Agricultural Development, www.ifad.org/media/events/2003/desertification.htm.

104. Corson, *Global Ecology Handbook*, 68, 77.

105. Lester R. Brown, "World Grain Production Falls to 57 Days of Consumption: Grain Prices Starting to Rise" (June 15, 2006), Earth Policy Institute, www.heatisonline.org/contentserver/objecthandlers/index.cfm?id=5973&method=full.

106. "Harvesting Poverty: The Rigged Trade Game," *New York Times*, July 20, 2003, www.nytimes.com/2003/07/20/opinion/20SUN1.html?ex=1233205200&en=12b0863506008c39&ei=5070.

107. Trainer, *Abandon Affluence!*, 153.

108. Wyn Grant, "Irish Farming Faces Big Shakeup" (March 29, 2006), Common Agricultural Policy, http://commonagpolicy.blogspot.com/2006_03_01_archive.html.

109. *Stop the Dumping! How EU Subsidies Are Damaging Livelihoods in the Developing World*, Oxfam Briefing Paper no. 31 (Washington, DC: Oxfam, 2002), 5, www.globalpolicy.org/socecon/trade/subsidies/2002/10stopdumping.pdf.

110. "Farm Subsidy Database," Environmental Working Group, http://farm.ewg.org/farm/progdetail.php?fips=00000&page=conc&progcode=total.

111. Joachim von Braun, "Overview of the World Food Situation—Food Security: New Risks and New Opportunities" (October 29, 2003), International Food Policy Research Institute, www.ifpri.org/pubs/speeches/20031029vonbraun.htm.

112. Trainer, *Abandon Affluence!*, 165.

113. Aditya Chakrabortty, "Biofuels Send Food Costs Soaring: Report," *Sydney Morning Herald*, July 5–6, 2008.

114. Trainer, *Abandon Affluence!*, 233.

115. Ibid., 173.

116. Walter Russell Mead, "UN Blessing Is Just a Frill for a US War in Iraq," *Los Angeles Times*, February 23, 2003.

117. Michelle Syverson, "GATT, the Environment and the Third World," *Environmental Law* 23, no. 2 (April 1993): 715–19.

118. Chakravarthi Raghavan, *Recolonization: GATT, the Uruguay Round & the Third World* (Penang, Malaysia: Third World Network, 1990), 37, 63.

119. Ibid., 91.

120. Ibid., 45.

121. "Expenditures for US Industrial R&D Continue to Increase in 2005: R&D Performance Geographically Concentrated" (September 2007), National Science Foundation, www.nsf.gov/statistics/infbrief/nsf07335.

122. Greg Clough and Ted Wheelwright, *Australia: A Client State* (Ringwood, Australia: Pelican, 1982), 4–15.
123. Raghavan, *Recolonization*, 98.
124. KRSNetwork, *2005 U.S. Pesticide Industry Report: Executive Summary* (Covington, GA: KRSNetwork, 2006), www.knowtify.net/2005USPestIndReptExecSum.pdf.
125. Vijesh V. Krishna, N. G. Byju, and S. Tamizheniyan, "Integrated Pest Management in Indian Agriculture: A Developing Economy Perspective" (May 25, 2007), Radcliffe's IPM World Textbook, http://ipmworld.umn.edu/chapters/Krishna.htm.
126. Syverson, "GATT, the Environment and the Third World."
127. Raghavan, *Recolonization*, 98.
128. Trainer, *Abandon Affluence!*
129. Barry Krissoff and John Wainio, "U.S. Fruit and Vegetable Imports Outpace Exports" (June 2005), USDA's Economic Research Service, www.ers.usda.gov/amberwaves/june05/findings/USFruitandVegetable.htm.
130. Syverson, "GATT, the Environment and the Third World."
131. "Ban on Tuna Imports Held to Violate Treaty," *Washington Post*, August 26, 1991.
132. Kanichi Ohmae, "Toward a Global Regionalism," *Wall Street Journal*, April 27, 1990.
133. World Trade Organization, www.wto.org.
134. Joseph Stiglitz, *Making Globalization Work* (New York: Norton, 2006), 79.
135. "WTO Sanctions against the US Are Urged," Associated Press, June 21, 2007.
136. Carter Dougherty, "Once Again, Trade Effort Stumbles on Subsidies," *New York Times*, June 22, 2007, www.nytimes.com/2007/06/22/business/worldbusiness/22trade.html.
137. Stiglitz, *Making Globalization Work*, 80.
138. Timothy Wise and Kevin Gallagher, "A Bad Deal All Round," *Guardian* (UK), July 30, 2008, www.guardian.co.uk/commentisfree/2008/jul/30/wto.economics.

9. The Manufacture of Consent

1. "Labor Day 2003 Toolkit. For Academics Who Support Workers Rights," AFL-CIO, www.aflcio.org/issues/jobeconomy/livingwages.
2. Alex Carey, "Managing Public Opinion"; Carey, "From Cominfirm to Capinform: A Bipartisan Approach to 1984"; Carey, "Conspiracy or Groundswell"; Carey, "Business Propaganda and Democracy." Copies of these unpublished works, written by Alex Carey between 1983 and 1987, can be obtained from Lou Kiefer, Western Regional Officer, International Machinists and Aerospace Workers, PO Box 1400, Oakland, CA 94604.
3. A. L. Lowell, *Public Opinion in Popular Government* (New York: Longman Green, 1926), 43.
4. George Brown Tindall, *America: A Narrative History*, 2nd ed. (New York: Norton, 1988), 990.
5. *The Penguin English Dictionary* (Harmondsworth, England: Penguin, 1985–86).
6. Tindall, *America*, 1005.
7. Alex Carey, *Taking the Risk Out of Democracy: Propaganda in the US and Australia* (Sydney, Australia: University of NSW Press, 1995).
8. Tindall, *America*, 1006.
9. Edward Bernays, *The Engineering of Consent* (Norman: University of Oklahoma Press, 1955).
10. H. D. Lasswell, *Propaganda Techniques in World War One* (Cambridge, MA: MIT Press, 1971), 222.
11. *Proceedings of the 40th Annual Convention of the Congress of American Industry* (National Association of Manufacturers, 1935), 25.
12. S. H. Walker and Paul Sklar, *Business Finds Its Voice: Management's Effort to Sell the Business Idea to the Public* (New York: Harper, 1938), 202.
13. K. Sward, "The Johnstown Steel Strike of 1937," in *Industrial Conflict*, eds. G. W. Hartman and T. Newcombe (New York: Corden, 1939), 74–102.
14. Richard S. Tedlow, "The National Association of Manufacturers and Public Relations during the New Deal," *Business History Review* 50 (1976), 25–45.

15. Daniel Bell, "Industrial Conflict and Public Opinion," in *Industrial Conflict*, eds. A. W. Kornhauser, R. Dubin, and A. M. Ross (New York: McGraw-Hill, 1954), 240–56.
16. Morris Bartel Schnapper, ed., *The Truman Program: Addresses and Messages* (Washington, DC: Public Affairs Press, 1968), 84–85.
17. Ibid.
18. Carey, *Taking the Risk Out of Democracy*.
19. Peter F. Drucker, *Have Employer Relations Had the Desired Effect?* American Management Association Personnel Services, no. 134 (New York: American Management Association, 1959).
20. Walter H. Corson, ed., *The Global Ecology Handbook: What You Can Do about the Environmental Crisis* (Boston: Beacon Press, 1990), 44.
21. Carey, *Taking the Risk Out of Democracy*.
22. Bradford Plumer, "How Rich People Control Politics," *New Republic* (January 30, 2007).
23. Ibid.
24. Mark Preston, "Political Television Advertising to Reach $3 Billion" (October 15, 2007), CNN.com, www.cnn.com/2007/POLITICS/10/15/ad.spending/index.html.
25. F. H. Knelman, *America, God and the Bomb* (Vancouver: New Star Books, 1987).
26. Cliff Kincaid, *The United Nations Debt: Who Owes Whom?* Cato Policy Analysis, no. 304 (Washington, DC: Cato Institute, 1998), www.cato.org/pubs/pas/pa-304.pdf.
27. Naomi Klein, *The Shock Doctrine: The Rise of Disaster Capitalism* (New York: Metropolitan Books, 2007).
28. A. Crittendon, "A New Corporate Activism in the US," *Australian Financial Review*, July 18, 1978.
29. Kim McQuaid, "The Round Table Getting Results in Washington," *Harvard Business Review* 59 (May–June 1981), 114–23.
30. Mark Green and Andrew Buchshaum, *The Corporate Lobbies: Political Profiles of the Business Roundtable and the Chamber of Commerce* (Washington, DC: Public Citizen, 1980), 15.
31. John S. Saloma, *Ominous Politics: The New Conservative Labyrinth* (New York: Hill and Wang, 1984), 14–15.

32. Carey, *Taking the Risk Out of Democracy.*
33. *Climate Change 2007: Impacts, Adaptation and Vulnerability: Contribution of Working Group II to the Fourth Assessment Report of the Intergovernmental Panel on Climate Change* (Cambridge: Cambridge University Press, 2007).
34. Peter Morici, "Dr. Peter Morici: US Records $178.5 Billion Third Quarter Current Account Deficit: Taxes Growth and Increases in Foreign Debt" (December 17, 2007), Finfacts, www.finfacts.com/irelandbusinessnews/publish/article_1012136.shtml.
35. Steve McGourty, "United States National Debt (1938 to Present): An Analysis of the Presidents Who Are Responsible for the Borrowing" (third revision, May 6, 2007), www.cedarcomm.com/~stevelm1/usdebt.htm.
36. Martin Mayer, "'A 'Family Man' Who Bled the Weak, Meek," *Sydney Morning Herald*, January 21, 1991, excerpted from Martin Mayer, *The Collapse of the Savings and Loan Industry* (New York: Scribner's, 1991).
37. Ibid.
38. John Durie, "US Banking Reforms Skirt Sensitive Issues," *Australian*, February 7, 1991.
39. Marcy Gordon, "U.S. Government Could Guarantee $2 Trillion for Banks" (October 14, 2008), Gold Anti-Trust Action Committee, www.gata.org/node/6783.
40. "US Treasury Urged to Examine Insurance," *Weekend Australian*, December 29–30, 1990.
41. Sheryl Canter, "Insurance Coverage Crumbles in Coastal States" (November 2, 2007), Yahoo! Green, http://green.yahoo.com/blog/climate411/53/insurance-coverage-crumbles-in-coastal-sates.html.
42. Rick Brooks and Constance Mitchell Ford, "The United States of Subprime: Data Show Bad Loans Permeate the Nation: Pain Could Last Years," *Wall Street Journal*, October 11, 2007.
43. Anatole Kaletsky, "The Real Reason Why Bankers Feel So Gloomy," *Times* (UK), July 17, 2008, www.timesonline.co.uk/tol/comment/columnists/anatole_kaletsky/article4346962.ece.
44. Ibid.
45. *The International Comparative Legal Guide to: Mergers & Acquisitions*

2007 (London: Global Legal Group, 2007), www.iclg.co.uk/khadmin/Publications/pdf/1130.pdf.

46. Martin Wolk, "Cost of Iraq War Could Surpass $1 Trillion" (March 17, 2006), MSNBC.com, www.msnbc.msn.com/id/11880954.
47. "Soaring US Debt Threatens Dollar" (December 26, 2007), US Economy Crisis: Recession 2007–2008, http://us-recession-2007.blogspot.com/2007/12/soaring-us-debt-threatens-dollar.html (accessed December 30, 2007).
48. Harry R. Weber, "Coca-Cola Reports Profit Rose 13 Percent in 3rd-Quarter As Sales Climbed" (October 17, 2007), *AP News*, www.blnz.com/news/2007/10/17/Coca-Cola_reports_profit_jump_2765.html.
49. Jason De Parle, "Crux of Tax Debate: Who Pays More?" *New York Times*, October 15, 1990.
50. Edmund L. Andrews, "Tax Cuts Offer Most for Very Rich, Study Says," *New York Times*, January 8, 2007.
51. Richard Rubin, "Most Corporations Don't Pay Taxes," *Congressional Quarterly*, August 15, 2008.
52. "War and Taxes: With 40% of IRS Revenue Going to Military, Resisters Prepare to Withhold Taxes to Protest War," *Democracy Now*, April 12, 2007.
53. John Wolper, "47 Million Americans Living without Health Insurance" (September 3, 2007), eFluxMedia, www.efluxmedia.com/news_47_Million_Americans_Living_without_Health_Insurance_08270.html.
54. Victoria Colliver, "We Spend Far More, but Our Health Care Is Falling Behind—Australia, Canada, Germany, New Zealand, U.K. Spend Less and Do Better Job, Studies Say," *San Francisco Chronicle*, July 10, 2007, www.sfgate.com/cgi-bin/article.cgi?f=/c/a/2007/07/10/MNGNUQTQJB1.DTL&hw=We+Spend+Far+More&sn=001&sc=1000.
55. Deborah Cameron, "Your Money or Your Life," *Sydney Morning Herald*, July 8, 1989.
56. James Arvantes, "Congress Provides Six-Month Reprieve from Medicare Payment Cuts" (December 19, 2007), American Academy of Family Physicians, www.aafp.org/online/en/home/publications/news/news-now/government-medicine/20071219medicarebill.html.

57. Patrick Purcell, "Income and Poverty among Older Americans in 2006" (September 2007), Digital Commons@ILR, http://digitalcommons.ilr.cornell.edu/key_workplace/312.
58. "Long-Term Growth of Medical Expenditures—Public and Private" (May 2005), ASPE Issue Brief, http://aspe.hhs.gov/health/MedicalExpenditures/index.shtml.
59. Henry A. Waxman, "Frenzy of Cuts in Medicine Plugs Deficit," *Los Angeles Times*, October 9, 1990.
60. Cameron, "Your Money or Your Life."
61. "Study: Geography Greek to Young Americans" (May 4, 2006), CNN.com, www.cnn.com/2006/EDUCATION/05/02/geog.test.
62. Michael Venture, "America by the Numbers. No. 1?" *Austin Chronicle*, February 23, 2005.
62. Steve Lohr, "With Finance Disgraced, Which Career Will Be King?" *New York Times*, April 11, 2009.

10. The Media and the Fate of the Earth

1. Brian Stelter, "A Lucrative Deal for Rush Limbaugh," *New York Times*, July 3, 2008.
2. Anup Shah, "Media Conglomerates, Mergers, Concentration of Ownership," Global Issues, www.globalissues.org/HumanRights/Media/Corporations/Owners.asp.
3. Amy Goodman and David Goodman, "Why Media Ownership Matters," *Seattle Times*, April 3, 2005.
4. Shah, "Media Conglomerates."
5. "Six Mega-Media Giants by Year 2000: Study," *Northern Star* (Lismore, Australia), February 2, 1991; Ben Bagdikian, "The 26 Corporations That Own Our Media," *Extra!*, June 1987.
6. *INFACT Brings GE to Light: General Electric, Shaping Nuclear Weapons Policies for Profits* (Boston: INFACT, 1988).
7. GreenBiz, "GE Ecomagination Revenues Cross $10 Billion" (May 19, 2006), ClimateBiz, www.climatebiz.com/news/2006/05/19/ge-ecomagination-revenues-cross-10-billion.
8. Doug Henwood, "Capital Cities/ABC: No. 2, and Trying Harder," *Extra!*, March–April 1990.

9. Peter Phillips, "Big Media Interlocks with Corporate America" (June 24, 2005), CommonDreams.org, www.commondreams.org/views05/0624-25.htm.

10. Peter Dykstra, "Polluters' PBS Penance," *Extra!*, May–June 1990.

11. "Study Aftermath: FAIR Debates MacNeil/Lehrer," *Extra!*, May–June 1990, www.fair.org/index.php?page=1531.

12. William Hoynes and David Croteau, "All the Usual Suspects, MacNeil/Lehrer and Nightline," *Extra!*, Winter 1990.

13. "Study Aftermath."

14. Ibid.

15. Ibid.

16. "Businessmen: TV's Oppressed Minority," *Extra!*, June 1987; Doug Henwood, "Public's TV's Elite Market," *Extra!*, Summer 1990.

17. Ibid.

18. "Who Is Rupert Murdoch?" (July 16, 2006), Center for American Progress, http://americanprogress.org/issues/2004/07/b122948.html.

19. Anthony Lewis, "Mr. Murdoch's Shadow," *New York Times*, November 5, 1987.

20. John Cassidy, "Annals of the Media: Murdoch's Game," *The New Yorker*, October 16, 2006.

21. Ibid.

22. David Barstow, "Behind TV Analysts, Pentagon's Hidden Hand," *New York Times*, April 20, 2008.

23. Mike Allen, "Exclusive: McClellan Whacks Bush, White House" (May 28, 2008), Politico, www.politico.com/news/stories/0508/10649.html.

24. Greg Clough and Ted Wheelwright, *Australia: A Client State* (Ringwood, Australia: Pelican, 1982), 33.

25. Jeremy Gerard, "Walter Cronkite: This Is the Way It Is," *International Herald Tribune*, January 10, 1989.

11. Healing the Planet: Love, Learn, Live, and Legislate

1. Adam Lioz and Gary Kalman, *The Wealth Primary: The Role of Big Money in the 2006 Congressional Primaries* (Washington, DC: U.S.

PIRG Education Fund, 2006), 1–7, www.uspirg.org/uploads/Wt/Fl/WtFlSC2GUfCSj1PiusxSGA/WeathPrimary2006.pdf.

2. Clara Bingham, "Queens of the Hill: Women Move to Leadership Positions in Congress," *Washington Monthly*, January 11, 2007, www.alternet.org/story/46485.
3. Arjun Makhijani, *Carbon-Free and Nuclear-Free: A Roadmap for U.S. Energy Policy* (Takoma Park, MD: IEER Press, 2007), www.ieer.org.
4. Ibid.
5. Ibid.
6. Matthew L. Wald, "Two Large Solar Plants Planned in California," *New York Times*, August 14, 2008.
7. Ibid.
8. Ibid.
9. Walter H. Corson, ed., *The Global Ecology Handbook: What You Can Do about the Environmental Crisis* (Boston: Beacon Press, 1990).

INDEX